Going to the Zoo
with Roger Caras

Going to the Zoo with Roger Caras

by ROGER CARAS

Illustrated by Cyrille R. Gentry

Foreword by William G. Conway, *Director, Bronx Zoo*

Harcourt Brace Jovanovich, Inc. / New York

Text copyright © 1973 by Roger Caras
Illustrations copyright © 1973 by Harcourt Brace Jovanovich, Inc.
All rights reserved. No part of this publication may be reproduced
or transmitted in any form or by any means, electronic or mechanical,
including photocopy, recording, or any information storage and
retrieval system, without permission in writing from the publisher.
ISBN 0-15-231130-0
Library of Congress Catalog Card Number: 72-88167
Printed in the United States of America
First edition
B C D E F G H I J K

For Pamela and Clay
who took me to the zoo many times
—with love

Foreword

With each passing day, wild animals become fewer and men become more plentiful. It is no wonder that zoos are attracting increasing interest. The zoo's existence and even its inner workings, so clearly set forth by Roger Caras, is an anachronism. Here twentieth-century men still maintain an interdependent relationship with wild creatures—the most prolonged and intimate relationship that man and wild animal have ever known. At the same time, the future of wild creatures in nature has become totally dependent upon man's interest and understanding. Zoos no longer exhibit live "samples" of dangerous unexplored wildernesses; instead they attempt to nurture tiny pockets of life from habitats that are rapidly disappearing.

The vast majority of people who visit zoos are city dwellers. Never again will they have the opportunity to see significant wildlife except in zoos and reserves; mostly in zoos. Yet these overwhelming populations of city dwellers will determine, through their interests or indifference, whether any wild places and wild animals are preserved, and in that fact lies the educational task of modern zoos. Mr. Caras has performed an important service in providing the kind of interpretive introduction to zoos and wildlife that can only result from deep concern and firsthand experience.

William G. Conway
General Director, New York Zoological Society
and Director, Bronx Zoo

The Wonders Our Eyes Can See

There is a very old story about a farmer who went to the zoo for the first time. He wandered around looking at animals that although strange to him looked something like animals he had known on his farm. The wolves looked rather like big dogs and the water buffalo looked something like his cattle. Porcupines he had seen before, and raccoons, of course, and even some of the antelope looked a little like goats—but then he came to the pen that held the giraffes.

The old man's jaw fell open, and he stared at the huge bull giraffe that towered above him. He looked at the bright markings of the animal's coat; he gazed at the enormous neck and the strange bumps on the bull's head and watched with wonder as the giraffe's long blue-gray tongue curled out and grasped some hay. Finally, they say, the man made up his mind. He shook his head and said, "There ain't no such animal." Satisfied with his decision, he moved along to pass judgment on the other animals.

But, you see, there was such an animal, even if the old man couldn't believe his own eyes. And the giraffe, as improbable as it is, however strange its design, is only one of the wonders that await the curious mind beyond the gates of the zoo. But, of course, you have to be prepared to believe what you see.

The Magic Gateway

Every time we visit a zoo we have a chance to pass through a series of magic gateways. An animal in an enclosure is not just an isolated creature sitting apart from the rest of the world. That animal is part of what we call a *system*. In its own land, living its own natural life, it would play a part in the lives of many other animals, and many other creatures would figure in its life as well. It might hunt or be hunted, or it might compete with others of its own kind for a mate. It might destroy and it might build, but it would be part of something bigger than itself, just as we are. The more we learn about an animal that we see in the zoo, then, the more we will know about wildlife and all of nature. And the more we know about nature, the more we understand ourselves. In a very real way, an animal in a zoo is a magic gateway opening up strange lands, strange creatures, and a whole new world of adventure for us to wonder at.

Why a Zoo?

Before we begin to understand the animals in a zoo we really should understand the zoo itself. It is quite a remarkable institution.

The idea of keeping wild animals in captivity for study and amusement is a very old one. The ancient Greeks and Romans did it two thousand years ago. But even earlier than that, the Chinese and Egyptians had zoos, perhaps as far back in history as four thousand years ago. For a long time it wasn't a very good idea. The men who kept animals understood little about them and didn't seem to care very much about their welfare. To them, a menagerie (that's the old name for a zoo—the word *zoo* is a shortened form of *zoological garden*) meant simply a collection of as many different animals as could be gathered in one place. For a time, in Rome, the captive animals were used in terrible displays. They were thrown together in an arena and forced to fight each other and armed men. It was then considered amusing to stage such spectacles. The animals were appreciated only for their strength and ability to fight.

But even the plain menagerie idea was a bad one. Animals were kept in small cages and suffered in a number of ways. Except for reptiles and only the very smallest mammals and birds, most animals should not be kept in small enclosures. They don't get enough exercise, the enclosures are difficult to keep clean, and the animals do not get a chance to behave naturally. It is not nearly as interesting for us to see animals in little steel boxes as it is in natural settings, so the menagerie idea is a bad one for people, too. It certainly can't help our appreciation of wildlife to see it in *jail*!

Perhaps the worst thing about the old menageries, though, was the terrible price wildlife had to pay to keep them supplied. In their little steel boxes, without scientific care, the animals didn't breed. They lived short lives, were thrown out, and replaced. Rare animals were made even more rare by hunters who kept the menageries supplied. The old menagerie was a cruel and wasteful idea.

But, hopefully, most of that is now history. The modern zoo is a very different kind of institution. There are four reasons for a modern zoo existing at all:

1. The zoo as an *educational institution*. As we said at the beginning, a zoo is a series of magic gateways. City and country people can gather there and learn about distant lands by coming to understand animals that are really *ambassadors* from those lands. In a zoo we can learn about conservation and ecology and the natural world. A zoo is a living museum, and, in a way, it is a school where the teachers are animals. Not everyone who goes to the zoo is like that old farmer. Some people not only believe the giraffe; they also learn from it.

2. The zoo as a *recreational facility*. A zoo is fun to go to, and more people go to zoos in America each year than go to all the baseball and football games combined. It is interesting, it makes our imaginations work overtime, and it is a change from our everyday life. A zoo is a place we can care about and enjoy and even help care for. In our modern world such places are becoming more important all the time. For people living in a city the zoo is a kind of bridge back to the wild and natural world. That, too, is very important. It helps us in many ways. The interests we develop make us want to read, to travel, and to protect what we have.

But those are two things that are easy for us to see and understand because they involve us directly. What about the things that go on behind the scenes? That is where we find the other two reasons for there being a zoo at all.

3. The zoo as a *research center*. A fine zoo brings together many different kinds of animals from many lands. It gives scientists a chance to study them at close quarters, to understand their behavior, to learn their needs. By studying their nutrition and their diseases, veterinarians and biologists can learn more about the things we must do in the wild places to help animals survive in our modern world. Without this understanding and concern, the wild creatures of the world will disappear. In some zoos in America now they are teaching chimpanzees how to talk in sign language, and they are discovering amazing things about animal intelligence—and about the origins of our own.

4. The modern zoo as a *survival center*. That may be the most important thing of all. Each year man destroys more of the wild habitats where wild animals live. Each year certain species of animals come closer to extinction. Unless we act very quickly, we could lose many of the wild animals we now take for granted. The great Indian rhinoceros and the tiger are good examples of animals that may vanish from the wild in the next twenty-five years. And that is where the modern zoo comes in.

In the old-fashioned menagerie, animals seldom bred. The people that ran them didn't try very hard to create the kind of homes for the animals that would encourage them to reproduce. Therefore, an animal brought into a menagerie was as good as dead as far as nature was concerned. But that is not the way it is done now. In modern zoos the emphasis is on breeding. Everything is done to make the animal feel natural and at home so that it will breed because for many kinds of animals the zoo is all that will be left in the years ahead.

There are already animals that have vanished from the wild. We have them today only because zoo men worked to save the few survivors and get them to breed in captivity. When we think of buffalo (really *bison*), we think of North America and the Great Plains. Well, there is a bison in Europe, too. It looks rather like our own and is called the *wisent*. There are no wild wisent at all left in the forests of Europe, but there are in zoos and in a few nature reserves where some zoo wisent have been placed—mostly in Poland.

In China, perhaps two thousand years ago, there was a rather strange-looking reddish gray deer the Chinese called *Mi-lu*. There were few left even then. Most had been killed off by peasants looking for food and by noblemen looking for something to hunt to relieve the boredom of their own lives. The few Mi-lu that managed to survive were gathered together into a game park by the emperor, and it became a crime to harm one. Since the emperor had the power to chop the head off anyone who disobeyed him, the Mi-lu were safe in his care. We call that deer *Pere David's deer* today, and there are no wild ones left anywhere—they are all in zoos.

Thousands of years ago a flock of Canada geese were blown off course by a great storm in the Pacific Ocean. They were migrating along the west coast of North America when the fierce storm struck. Weeks later a few stragglers ended up on some volcanic islands. It was a miracle that any survived. Hungry, bruised, exhausted, most died, but a few somehow made it. In time they began to breed, and they never left their new island home again. The islands where they landed and remained later became known as the Sandwich Islands and then by their rightful name the Hawaiian Islands. The geese had arrived there before man, but, in time, the islands were discovered and the geese suffered as each new wave of human settlers came ashore. They were hunted for food and sport, their eggs were collected and eaten, and dogs, cats, and mongooses brought by the settlers to hunt rats preyed on their chicks. The pigs and goats that came with the settlers turned out to be enemies, too, because they trampled the nests of these ground-nesting birds.

Today we call that goose the Hawaiian goose or *nene*—that is pronounced *nay-nay*. Just a few years back there were not even a hundred nene left out of the original flocks that had numbered in the thousands. Citizens of Hawaii, concerned over the fate of their handsome goose, captured those few that remained and shipped them halfway around the world, to southern England, to a special zoo called *The Wildfowl Trust*. There, a conservationist named Peter Scott keeps almost all of the *species* of ducks and geese there are and breeds them. Working with the nene that had come to him from the distant Pacific, Peter Scott and his scientists set up pens and ponds where the geese could feel at home. In time they began to breed again, and soon there were enough nene so that breeding pairs could be sent off to other zoos. And now, with the captive flocks dispersed around the world, nene are being returned to Hawaii and turned loose to take their rightful place back in the wilds again. The adventure story of those storm-tossed Canada geese of thousands of years ago and their descendents has a happy ending because of the modern zoo.

That, really, is what the zoo today is all about. It teaches us how to understand animals, it gives us an opportunity to enjoy animals while we learn about them, and it gives the animals a chance to survive. In Phoenix, Arizona, there is a herd of rare and beautiful antelope called *Arabian oryx*. When it became apparent that the oryx could not survive in its native Arabia because of hunters that could not be controlled, specimens were captured and shipped to the American Southwest, an area that somewhat resembles their own habitat. What was done with the nene on the banks of the Severn River in England will be done with the oryx in Arizona. In Israel they are saving a rare and beautiful wild goat called the *Nubian ibex*. The *Mongolian wild horse* is surviving in Prague, Czechoslovakia, and at the Catskill Game Farm in New York. All over the world modern zoos that in no way resemble their ancient counterpart—the menagerie—are working to make the world safe for wildlife.

Now, let's see how a zoo works.

What It Takes to Make a Zoo Work

It takes people to make a zoo work, a great many people with special talents. The men at the top are called directors or *curators*, and their job is a difficult one. Since millions of people may visit a big city zoo in a single year, the director of the zoo must be concerned with their safety and comfort. There are problems of traffic, parking, sanitary facilities, food concessions. The same man at the top has to worry about the dozens of men and women who work for the zoo organization. Their needs and welfare must be provided for, so in every sense a zoo is a huge business enterprise. To the visitor it may look like a park (if it does, the zoo men there are skillful), but in fact it is a far bigger business than many of the factories and offices that surround it. But those are things you never see. They are the hidden world of the zoo.

It takes food to run a zoo—tons of it and of many different kinds. It must be fresh, it must be clean, and it must be tested to be certain harmful chemicals haven't contaminated it. Most animals eat every day, and every day, no matter what the weather, no matter what the human problems, the food must be prepared and distributed.

Not all the foods used are easily obtained. Some of them may have to come from far away, and the mixtures are often complex. Animals have special needs, and if the animals are to survive and hopefully breed, those needs must be met at all costs.

What does a zoo menu look like? Let us say you had the smallest zoo in the world. Let us imagine you had one elephant, one hippopotamus, one lion, one gorilla, one woolly monkey, one walrus, one koala bear, one grizzly bear, one polar bear, and one zebra. What kind of food would you need every day? (Actually, of course, it is very bad zoo practice to keep one of anything. Zoo directors always try to have *at least a pair* of everything they exhibit—in the hopes that young will appear.) This is what you would need for your mini-zoo *every day* of the week:

800 pounds of alfalfa and timothy hay
 16 quarts of ground grain (with minerals and salt added)
 25 loaves of bread
 10 cabbages
  4 quarts of crushed oats
 25 pounds of raw white potatoes (some whole and unpeeled, some whole and peeled, and some diced or chopped)
 30 pounds of apples and carrots (some whole, some sliced, some chopped)
 12 pounds of pellets made from various grains, minerals, and vitamins
 12 ounces of cream of wheat
  1 quart of white milk
  1 quart of malted milk
  5 cans of evaporated milk
  ½ cup of honey
  ½ quart of corn oil
  ½ pint of cod-liver oil
  1 quart of ground-up fish
 40 pounds of mackerel—whole
 20 pounds of other fish (butterfish, etc.)—whole
  6 raw eggs
  1 hard-boiled egg
  3 slices of Zwieback with currant jelly
  1 pint of Jello
  6 stalks of celery
  1 head of lettuce
  1 handful of grapes
  1 quart of weak tea (with raw eggs in it)
 10 cookies or wafers
  5 ounces of canned dog food
  1 whole orange
  8 ounces of sliced peaches with pineapple and strawberries

    5 sliced plums or apricots
    2 bananas
    1 cup of diced fresh mixed vegetables and greens
    1 cup of diced fruit
  30 pounds of raw meat (with bone meal and cod-liver oil added)
      eucalyptus leaves (fresh, mature leaves only)
      salt block with iodine
      multiple vitamin concentrate in sweetened orange juice

All this food plus several thousand gallons of clean, fresh drinking and bathing water every day—and remember, your mini-zoo contained only ten animals, and that did not include a single bird or reptile. Compare your mini-zoo with one like the Bronx Zoo in New York, where there were recently 841 mammals, 2,005 birds, and 544 reptiles and amphibians. The "appetite" of the Bronx Zoo would be 339 times as great as that of your ten-animal mini-zoo. As you can see, it must take a lot of food to run a zoo even for a day.

But the food itself is not the whole story. There have to be people who know how to order the food and prepare it and, most importantly, ration it. The animals must be watched to be certain that they are eating what they get and getting what they need. That means teamwork between animal *nutritionists* and *veterinarians*. And when you come to the veterinarian, you come to one of the hardest working and most highly trained persons in the zoo. In a single day a veterinarian may help a zebra deliver her foal, extract an infected tooth for a very angry tiger (and angry tigers are certain to be unpleasant patients), give a gorilla a vitamin shot, file an elephant's toenails, evaluate the diet of a newly arrived pygmy hippopotamus, set the broken arm of a spider monkey, then reset the dislocated leg of a gazelle. If he is lucky, he will not be called upon in the middle of the night to help a rhinoceros who is having a stomachache. One thing is certainly true, the life of a zoo veterinarian is never dull. And despite all of his hard work, none of his patients every say "thank you."

Once upon a time the zoo veterinarian had to rely on brute force to subdue an angry or frightened wild animal. Good zoo people rarely resort to that kind of thing now, for they know it upsets the animals in their care. Animals are given drugs in their food or shot with an almost painless dart gun that injects a *tranquilizer* into their bloodstreams. These drugs make the animals feel drowsy and enable the veterinarian and his helpers to work on the animals without hurting them or getting hurt themselves.

There can be times when the veterinarian would rather not give drugs to an animal (when it is a female carrying unborn babies inside of her, for example), and then a squeeze-cage is used. An animal is lured into a special cage. When the door is shut, the keepers turn cranks that move two walls inward on tracks set in the cage floor and ceiling. In short order the animal is *immobilized* so that so that it cannot bite or strike out at the veterinarian as he reaches through the bars to examine an injury or administer a medicine.

But there is even more. A good zoo has a *pathology* section. In every zoo, of course, animals die of old age. These animals are given an *autopsy* by which the veterinarians and zoologists can learn new facts about disease, anatomy, and the other secrets hidden inside of living animals. Even in death a zoo animal can serve its own species by helping the people who have dedicated their lives to the welfare of wild creatures.

Zoo Orphans

Unfortunately, not all animal mothers are interested in their young. Captivity seems to upset some, and although they will breed in the zoo, they will not perform a mother's duties. There have been cases where females have killed or injured their young. That is where the zoo nurses must step in.

Mammals that the zoo director and his staff veterinarian feel may be neglected are taken from their mothers and turned over to especially trained girls who have the patience and love of animals necessary to raise them from infancy. The babies must be fed very often, they must be kept warm and clean, and they must be given both love and attention. If they were being raised by their natural mothers, they would be cuddled and constantly cared for, and this is necessary for their welfare. The nurses must substitute handling and then, eventually, play. It is a full-time job, and it becomes more strenuous every day as the young animals begin to move around with greater ease. Very soon they are capable of getting into trouble and must be protected from their own foolishness. The baby mammals become very attached to their foster mothers, and they probably don't even realize that they are not their natural ones. When a baby animal thinks a human being is its mother, we say that the animal has been *imprinted*.

The nurse's job is most difficult, perhaps, because she knows that none of her babies can remain as pets. In every case the day must come when the animal is returned to the normal life of a zoo animal. Many nurses find it very hard to give up the babies they have cared for with such love and attention for many, many months. More than one secret tear has been shed by a nurse as she said good-bye to her lion, tiger, gazelle, or even rhinoceros baby.

Is the Zoo a Dangerous Place to Work?

The stories of zoos down through the years are filled with wonderful friendships between men and animals. A keeper in the Bronx Zoo in New York used to sit in the lap of his favorite charge, a gorilla that weighed twice as much as he did. Sue Pressman who worked at the San Diego Zoo and later at the Franklin Park Zoo in Boston raised over one hundred and fifty lions, tigers, leopards, jaguars, cheetahs, and mountain lions in her own home. A keeper in the zoo in San Francisco used to treat his chimpanzee charge like his own child. The two of them, man and ape, could be seen on any day walking through the zoo grounds hand-in-hand.

As nice as these stories are, they are the exceptions. The zoo is a home for *wild* animals, and wild animals don't always act reliably around people. It is not so much a case of an animal wanting to hurt people—it is more often a case of an animal not knowing his own weight or strength. In a zoo where I once worked, we had a hippopotamus who weighed 5,400 pounds. She loved her keepers, and she seemed to love me when I went into her enclosure to visit with her. Very often I carried lumps of sugar with me and would give them to her as a treat. She would come up out of her pool and follow me around the enclosure. It was a bad habit to encourage. One day she came up behind me and began to nuzzle me asking for her sugar. I turned to face her and was scratching her on the nose, as I usually did when she approached, when I suddenly realized I was in serious trouble. Without meaning me any harm, Cuddles had backed me up against a stone wall and was pushing me with her nose. I was just barely able to slip sideways in time. Another inch or two and I would have been pinned. Without even knowing what she was doing, she would have broken my back and left me crippled or dead.

Because of the almost unbelievable strength of wild animals, most zoos today have rules about men exposing themselves to danger. It can be frustrating for the people who work there because it is possible to become extremely fond of an animal and to know by the way it acts how fond it is of you. This is particularly true of animals that were raised by nurses. Still, most zoos require their keepers to satisfy their feelings of affection for their animals by scratching them from behind a protective barrier. Very few allow the keepers to enter enclosures with dangerous animals. The amount of danger a keeper is exposed to, then, usually depends on how good he is at observing the rules.

You and the Zoo

We are going to discuss the animals you may expect to see in the zoo. Before we do, though, we should discuss one rather special creature—you. You have a role to play when you visit the zoo. That role is the observation of *zoo manners*. What are the rules of good zoo-visitor behavior?

1. Observe the rules about feeding animals. Most animals should *not* be fed; some may be fed, but only the special foods you can buy at the zoo. Follow the posted instructions exactly. Good zoo-keeping means careful diet control. Help your zoo maintain that control. When you buy food from the zoo to feed the animals, you are contributing to the care the animals receive. That makes you a part of the zoo. And that can be a good feeling.

2. *Never*, under any circumstances, throw anything at the animals or anything into their enclosures. Things thrown into enclosures are sometimes eaten by the animals, and animals have been killed that way. In one zoo a rubber ball was thrown to a seal by an unthinking boy. It seemed an innocent-enough thing to do, but the seal swallowed it and died. In another case a beautiful mountain lion was very nearly killed by a small rubber toy it had swallowed. If you see anyone throw anything at an animal (to wake it up or get it to move) or throw anything into an enclosure, report the incident to a keeper. He may have to remove the thrown object before it harms the animals. Help protect the citizens of your zoo from harm.

3. *Never*, under any circumstances, cross a boundary fence or a safety barrier. It is easy for an animal to misunderstand your

intent, and if you are injured, it could become necessary to destroy a fine, rare animal. Follow the rules and never attempt to touch an animal except in supervised petting or contact areas. When you are in areas where you are allowed to touch the animals, remember they are playmates and not playthings. Be careful, be gentle, be very quiet, and follow instructions. The animals are there so you can get to know them, not so you can terrify them. Be kind.

4. Noise is distracting to everyone, animals and people. While it is true that a zoo is a place where you go to have a good time, that doesn't mean that you can give other people a bad time. Animals are made nervous by screaming people, and that could interfere with their reproducing. Don't run around crashing into people and things; don't upset the animals. Be quiet and you will not only have a good time—you will also learn about animals and nature. (For goodness' sake, *don't* take a portable radio to the zoo with you!)

5. All good zoos have rules against pets. There is a very important reason for that. Not all pet animals are as carefully cared for as zoo animals, and the zoo veterinarian does not want animals coming in that could carry *communicable diseases*. There have been cases where pets got loose and either injured zoo animals or were injured by them. Observe the "no pet in the zoo" rule.

6. One of the biggest problems the zoo has to face is "housekeeping." That means keeping the animal areas clean, and the human areas, too. Very often the human areas are much harder to deal with than the animal enclosures. If you throw paper and other trash on the ground, a member of the zoo staff has to pick it up. If he is picking up after you, he is not caring for an animal. He can't do both at the same time. When you litter in a zoo, you are harming the animals. Don't do it. As a matter of fact, help. If you see something on the ground, pick it up and put it in a trashbasket. There are plenty around in every zoo.

7. A zoo is designed to be pleasant for everyone. Zoo directors

and their staffs try to have flowers and trees growing around the grounds for the pleasure of the human visitors and for the added welfare of the animals. Plants help reduce noise, they help keep the air fresh and pleasant smelling, and they create shade and help keep the temperature stable. They also block wind. Don't *ever* pick or injure flowers or shrubs, and never injure or scar a tree. They are living parts of the zoo, too. Remember, a zoo is about nature—and nature includes wild animals, plants, and you!

8. A zoo director has only so much money to spend each year. Sometimes unthinking visitors force the zoo director to take the money away from the animals and spend it on them instead. How? By injuring property—buildings, signs, waste receptacles, benches, fences, anything that might need replacement or repair because some unthinking person injured it. Treat the zoo as if it were your own private property and really care for it. (There is no other way you *can* act if you like animals.)

9. Many zoos have special volunteer programs, and you might want to ask about them. The day may come when you will be able to help by being a junior keeper or by guiding smaller kids around.

10. Many zoos today have membership programs. Some publish excellent animal magazines with not only news of that zoo, but of the world of animals in general. Become a member of your zoo's society and help support its terribly important work.

When you go to the zoo, there are projects that can help make your visits more valuable, and more fun, too. Here are some ideas:

TAKE PICTURES. Take your camera to the zoo. In order to photograph the animals you will see there in their *own* homes, you would have to travel completely around the world, first on an east-west line and then on a north-south one. It would take you years, and you still might not succeed in getting them all. *The zoo is one of the greatest photographic opportunities on our planet.* Keep a zoo scrapbook and make notes alongside the photographs. Identify the animal in the picture, note where it comes from, what kind of habitat it lives in (desert, jungle, plains). Note what it was doing when you saw it; if it had young; how old it was; the keeper will tell you if the sign doesn't. A zoo scrapbook can help you really understand the animals you are seeing and the role they play in nature.

TAKE RECORDINGS. Zoo animals, most of them at least, have voices, and they call to each other. Take a tape recorder along and record the sounds they make. Before or after each animal recording, record your own voice telling what the animal is and something about it. Play these recordings for your class. You will probably have more luck with the birds than you will the mammals (except the monkeys, of course). Except for a rare bit of luck when an alligators roars, the reptiles will probably be very poor hunting for your tape recorder. There is a great game you can play. Instead of recording your own voice explaining what the sounds on the tape are, make notes. Then get your family and friends to *guess* the identity of the animal voices they hear. (If you are ever lucky enough to get the voice of an ostrich, your friends will really be baffled. An ostrich sounds like a lion.)

TAKE YOUR SKETCHBOOK ALONG. You may be surprised by how good you are at making drawings of the animals you see. Even if your first efforts aren't great works of art, keep at it. A zoo sketchbook can be a thing of pride for you and your family. Be sure to date each sketch you make and carefully identify the subject.

LEARN GEOGRAPHY. Animals are a living geography lesson. Pin up a world map in your bedroom or in your classroom. Make neat little tags with the names of the animals you have seen and paste them on the map in their proper homes. You may want to include photographs you have taken. See how long it takes you to identify and photograph animals from all continents for your map project.

CREATE CHARTS. As a class project, or a project you do on

your own for your class, create charts out of the pictures you take (or get from magazines) and the things you learn. Show the five vertebrate groups—mammals, birds, reptiles, amphibians, and fish —and show the differences and the ways in which they are alike. Do a chart on nutrition. Show what an ostrich eats, a bear, a walrus, and the other zoo animals.

SEEK MAN'S HISTORY. Make a chart or a notebook showing the ways in which animals have helped man, and use the animals in your zoo for examples. There are many that are easy to see: the camel, of course, is called the "ship of the desert"; the elephant does the heavy work in Asia; the rare Mongolian or Przewalski's horse you may see in your zoo is closely related to the *ancestral* horse of all those we ride and use today. The wild red jungle fowl is where our chickens came from; the wild turkey gave us our domestic bird; from the wolves we got our domestic dog, or at least a great many of them. There may be wild geese and hogs in our zoo. There are lots of examples. We would not be where we are today in our civilizations if it were not for the animal kingdom. The zoo is a fine place to come to understand that debt.

Should You Have Your Own Zoo?

There is a temptation after you come home from the zoo to want one of your own, but there are some very serious problems. Wild animals don't make good pets. Some pet shops sell monkeys, ocelots —animals like that—but what they are doing is really dishonest and cruel. The animals they are selling should be back in the wild where they can reproduce, for they won't do it in your home. Those animals suffer, too. Pet shops and private homes cannot provide the kind of care they need. They need special foods and special medical care. Ask your veterinarian what he thinks, and you will learn soon enough what a bad idea exotic pets are.

Are there any wild pets that are good for the home zoos? Yes, some. Hamsters, gerbils, and the other small-cage rodents are descended from wild animals, but they are now bred in captivity and make satisfactory pets. There is a big advantage to them—they breed regularly, and you can have young to observe and even trade with friends. Parrakeets, lovebirds, and cockatiels are also now captive-bred and may be kept. Fish and other normal aquarium animals are fine pets, and many of them also breed in the home. You can have a terrarium and keep local frogs in it. The science teacher in your school can tell you where you can order *oothecas,* egg cases for the splendid insect we call the praying mantis. You can hatch the babies out by following instructions and then turn them loose to control insect pests. (You have to turn them loose after they hatch or they will eat each other!)

If you are very good at keeping your home zoo, you can even keep small local snakes and lizards. Before you take any turtles into captivity, though, write to your State Department of Conservation, of the Fish and Wildlife Bureau in your home state, and get a list of *protected species.* You can catch salamanders and interesting insects. There are all kinds of terrarium projects. *Never,* under any circumstances, however, think of keeping *venomous* snakes in the home. Zoos have very elaborate procedures to protect the keepers from venomous snakes, and even then accidents happen. Zoos keep special medicines (*antivenins*) on hand in case of accidents. These are things that are difficult to do in the home. *No* venomous snake makes a safe pet. They are all quite cranky and very, very fast!

So, there are animals that can live in a home zoo. Just follow this list of very important "Don'ts":

Don't keep animals that are protected by law. Remember, you are the animals' friend, not their enemy.

Don't try to keep venomous animals.

Don't keep exotic animals that have not been bred in captivity as pets. Don't encourage pet shops to sell them by being a cash customer. (If you see a pet shop selling animals like monkeys and ocelots, tell the owner you will do your business elsewhere until he gets rid of the animals he shouldn't have.) And that important *don't* includes parrots and macaws that have to be removed from the wild.

Don't keep animals you cannot properly care for. For instance, don't keep a snake unless your family is happy about it. Snakes need constant temperature, between 72 and 80 degrees, usually. If someone in your family is unhappy about a snake and you have to keep it outside or in a garage where you cannot control the temperature, then perhaps you shouldn't have a snake at all.

Don't obtain any animal until you fully understand its diet and other needs. Only when you really know what that animal must have to be healthy and *know* that you can provide it 365 days of the year should you obtain the animal itself. Remember, being a zoo keeper is not a sometimes job. The animals' needs are there every day without fail.

The Animals in the Zoo

Animals are what a zoo is all about. So let's discuss the animals you can expect to find. The more you know about them, the more you will enjoy seeing them.

The word *animal* itself seems to confuse some people. How often have you heard people speak of *birds and animals*? To speak of *birds and animals* is like speaking of *people and human beings*. Birds *are* animals, just as people are human beings. There are three different *categories* into which everything you can see in a zoo will fall. The masonry out of which the buildings are made, the rocks in the ground and the pavement—those are *minerals*. The trees, bushes, and flowers, and some of the food you will see in the animals' cages, are *vegetable matter*. Everything that is not mineral or vegetable that you will see will be animal. A bird and a fish are as truly animals as lions and giraffes are. A snake is an animal and so is an ant. But there are many different kinds of animals. Animals without backbones are called *invertebrates* and include insects and worms, lobsters, clams, spiders (which are *not* insects), and many more. There are many times as many *invertebrates* as *vertebrates*, but they are not very often found in zoos. *Vertebrates*, or animals with backbones, are divided into five groups: mammals, birds, reptiles, amphibians, and fish. We will discuss mammals first.

The Mammals

Mammals are the highest form of animal life. Mammals and birds both had reptiles as their ancestors, but there have been true mammals since the age of dinosaurs, 185 million years ago. If *all* of the following things are true about the animal you are looking at, it is a mammal:

1. It must breathe air to live. Even whales, seals, sea lions, and walruses—which are all mammals—must come to the surface of the water to fill their oversized lungs with air.

2. It has at least some hair or fur at some stage of its life.

3. It has a backbone and an internal skeleton.

4. The mother has special glands called *mammaries* (that is why we call them mammals) in which she produces milk. The baby animals feed on this milk—they nurse—until they are old enough to eat the kinds of food their parents eat. When the day comes that the baby animal no longer needs its mother's milk, we say that animal has been weaned.

5. The animal has a constant internal temperature. We refer to both mammals and birds (but *no* other animals) as being *warmblooded*. They have the power to produce their own heat and regulate it. That is why birds and mammals can live in places where most other animals cannot—places like the Arctic, near the North Pole.

6. One last thing. Although you cannot see it by looking at the animal itself, the mammals have the largest and most complicated

brains in the animal kingdom—they are the most intelligent of animals. (We are mammals.)

That is the mammal—the kind of creature some people think of as the only animal. As we know, they are mistaken. The second highest group of animals consists of birds.

The Birds

What makes a bird?

1. Like mammals, all birds must breathe air.

2. But there is an even simpler way to tell. If the animal you are looking at has feathers, it has to be a bird. If it does not have them, it cannot be a bird. All birds have feathers. No other animals do. (One hundred and forty million years ago there was a flying reptile that had feathers. But it has been long extinct, and its descendants are probably our birds.)

3. The birds, again like the mammals, have backbones and internal skeletons.

4. The birds, as we have said, are the only warm-blooded animals in the world besides the mammals.

5. All birds lay eggs. Only two mammals do (the duck-billed platypus and the spiny anteater from Australia and New Guinea).

6. Once again you can't see it by just looking, but the birds are the second most intelligent or advanced group of animals on our planet.

People often think of flight when they think of birds, and that is natural. But not all birds can fly. Many of the very large birds—the ostrich, emu, rhea, and cassowary—cannot, and the kiwi, the national bird of New Zealand, like many New Zealand birds, can't fly either. The large flightless birds are called *ratites*, and there is an interesting reason why. Birds that fly have powerful muscles to work their wings for hours on end—and often at great speeds. (Some hummingbirds beat their wings as fast as two hundred times a second!) The muscles have to attach someplace, and nature designed a ridge or keel in the middle of the flying bird's breastbone or *sternum*. The flight muscles on each side stretch across half of the bird's chest and connect to the ridge in the middle. Flightless birds do not have those powerful muscles, and so there is no need for that keel in the middle of the chest. The chest bone in the flightless birds is as flat as a raft. The Latin word for raft is *ratis*, so the birds are called *ratites*. That is how many scientific words have come into being.

The Reptiles

Once we leave the mammals and the birds and come farther down the scale in the world of animals, we come to the first *cold-blooded* animals—the reptiles. Cold-blooded is often used to mean cruel or unfeeling, but when we speak of animals, that is not what it means at all. It simply means that the animal doesn't have the delicate mechanisms necessary to keep its temperature the same *inside* no matter what happens to the weather *outside*. (If you like really big words to surprise your parents and friends with, here is a good one: the scientific word for cold-blooded is *poikilothermic*. Warm-blooded is *homoiothermic*.) All animals other than birds and mammals are pretty much the same temperature as the air or water around them. That single fact controls the way they are able to live.

What makes a reptile?

1. All reptiles have an internal skeleton the way mammals and birds do. Even the turtle, with his shell on the outside, has a backbone under it.

2. No reptile has either fur or feathers. All have *scales*. On some reptiles these scales are large and easy to see. In others they are so fine that you must look closely to find them.

3. Some reptiles lay eggs while others give birth to living young. Reptile mothers do not feed or really care for their young. When they are born or hatch, they are ready to care for themselves.

4. Snakes and some few lizards do not have legs—most lizards and all other reptiles do.

5. Reptiles are not really very smart. Although they are the third highest group of animals, their brains are rather small and not very advanced.

6. All reptiles need air at every stage of their lives. No reptile can breathe underwater.

The Amphibians

The fourth group of animals you can expect to see at the zoo consists of a small and rather ancient class called the *amphibians*. The fish were the earliest animals with a backbone, but they were held prisoner by the world of water in which they lived. In order for higher animals to invade the land, there had to be an intermediate group—something part way between the fishes and the reptiles—and that group turned out to be the *amphibians*.

The amphibians are confused with reptiles by some people. They are easy enough to tell apart. They are a small group when compared to the others—just the frogs, toads, and a small group of wormlike animals called caecilians, various newts and salamanders—and they are really not like the reptiles.

1. Amphibians also have a backbone and internal skeleton.

2. Amphibians do not have fur or hair or feathers, or even scales. Their skin is slimy. (Snakes and lizards, scaled reptiles, do not have slimy skins. Their skin is cool and dry and pleasant to touch.) Another way to tell the amphibians from the reptiles is that reptiles have claws while most amphibians do not.

3. Almost all amphibians lay eggs, but most of the time—although there are exceptions—the parents pay little attention to the eggs or to the young once they are born.

4. Most amphibians can breathe underwater for at least part of their lives. Amphibians show that they are descended from fish since most must return to the world of water to reproduce. Almost all

amphibians lay their eggs in water, and that is where the young hatch out. Tadpoles, for example, are baby frogs, but they usually live in water like some strange little fish. When they mature, they come up onto the land and breathe air as the higher animals do.

5. We do not think of amphibians as being very intelligent. While this is true, we must remember that they have been smart enough to survive for hundreds of millions of years. They seem to have just enough intelligence to make it through a hostile and dangerous world.

6. Since amphibians have no means of controlling their own temperature, they are, like reptiles, limited to temperate and tropical parts of the world.

There, then, are the four groups of backboned animals we can see in our visit to a zoo: the mammals, the birds, the reptiles, and the amphibians. Some zoos have representatives of the fifth group—the fish. But fish are usually kept in institutions known as aquaria or aquariums. They are a special subject all to themselves.

Before going on to talk about the animals in each group we may see, we should settle one very troublesome matter—at least some people find it troublesome—and that is the matter of scientific names. On the signs identifying the enclosures of the various animals, you will often see names you cannot pronounce, and you may wonder why anyone would want to bother with them at all. For instance, on the enclosure that holds the African elephant you will see *Loxodonta africana*. On the pen that holds the white rhinoceros you will see *Ceratotherium simum*. "Why bother with words like that?" a great many people ask. There is a very good reason why we do.

Around the world thousands of languages and dialects are spoken. In each of these languages there are names for animals. Not all scientists speak English, French, German, Spanish, Italian, Chinese, or Russian—not by any means. How would a scientist who spoke Hindi, Urdu, Bengali, or Serbo-Croatian know what you were talking about if you were a scientist in America, Canada, or England

and wrote about a white rhinoceros? He wouldn't. You wouldn't be able to communicate with him, especially if what you had to say was in writing and you weren't nearby so that he could ask you questions. That is one reason why every animal and plant in the world has to have a name that stays the same in every country where the animal is discussed. There is another reason, too. So-called *common names* change from place to place and are given to quite different animals. A good example is the elephant. If you are in India, *elephant* means an animal known to scientists as *Elephas maximus*, while in Africa it means the *Loxodonta africana* we were talking about before. As another example, there is a snake in the United States called the *copperhead*. There is a snake in Australia called the *copperhead*. They have nothing whatsoever to do with each other. They aren't even related. If scientists didn't have scientific names to go by (*Agkistrodon contortrix* for the American version and *Denisonia superba* for the Australian), it could get hopelessly confusing. Since there are more than a million different kinds of animals now known (including all the insects), there has to be some way of naming them. So science has worked out a system of names from Latin and Greek words, and they often appear on zoo cages and enclosures. You don't have to read them if you don't want to, but you can if you do. We are going to do the same thing in this book. We are going to use the common *and* the scientific names. Read them if you want to—ignore them if you don't.

Whenever you see the sign below in this book or in the zoo, it means the animal is an *endangered* one—one that will disappear unless we do everything in our power to protect those that are left and the habitat they live in.

Now on to the animals in the zoo.

1 / Virginia Opossum (*Didelphis marsupialis virginiana*)
A relative of the kangaroo, the Virginia opossum is the only one of the primitive mammals known as marsupials living in all of North America. It is neither very bright nor very fast and gets along by being active only at night. And it is true that these animals do play possum when they are threatened, once they decide hissing and growling isn't going to help.

2 / Koala Bear (*Phascolarctos cinereus*)
The koala bear isn't a bear at all but another kind of marsupial from Australia, where most marsupials come from. Koala bears have a very specialized diet and can eat only the old stale leaves of a few kinds of gum or eucalyptus trees. Only the largest zoos have them, for they are hard to keep.

3 / European Hedgehog (*Erinaceus europaeus*)
The hedgehog isn't a hog any more than a koala bear is a bear. It is a kind of insectivore, which, as the name implies, means it eats insects. It is a shy creature and moves about at night. When threatened, the hedgehog rolls up into a ball and presents a spiny exterior to the world.

4 / Red Kangaroo (*Macropus rufus*)
The red kangaroo inhabits the inland plains that cover most of Australia. When the kangaroo leaps, its tail serves as a balance and rudder, and when the animal is sitting, the tail is a third leg. It is strong enough to support the kangaroo's entire weight.

5 / Duck-billed Platypus (*Ornithorhynchus anatinus*)
The platypus, along with the echidna also pictured on the opposite page, belongs to the most primitive group of mammals left on earth, the monotremes. These are the only mammals that lay eggs. The platypus is found in Australia's rivers and nowhere else in the world. It can be seen in only the largest zoos.

6 / The Spiny Anteater or Echidna (*Tachyglossus* sp.)
There are five kinds of echidnas found in Australia and New Guinea. They are seen in zoos more often than the platypus and seem to get along pretty well on a diet of milk, beaten eggs, and shredded meat.

1 / Golden Horseshoe Bat   (*Rhinonicteris aurantia*)
Many zoos exhibit one kind of bat or another—usually in darkened rooms with only a red light glowing. Bats, being noctural, are blind to the color red and will move around in such a light, thinking it is night. They find their way by squeaking very rapidly and guiding themselves by the echoes. This species comes from Australia.

2 / Woolly Monkey   (*Lagothrix lagotricha*)
This forest monkey of South America is commonly seen in zoos. It is a hardy, good-natured animal and does well when given good care and a well-rounded diet. You can tell it is a New World monkey because it has a prehensile tail—a tail it can use like an extra hand—which African and Asian monkeys do not have.

3 / Pygmy Marmoset   (*Cebuella pygmaea*)
The marmosets of South America are the smallest of all the monkeys, and this species is the smallest of the marmosets. The marmosets are active and nervous little animals who may become extinct if great care is not taken to protect them in their forest homes and to breed them in our better zoos.

VANISHING ANIMAL

4 / Kirk's Red Colobus   (*Colobus badius kirkii*)
The colobi are among the most handsome of all the monkeys. The black and white colobus is more commonly seen in zoos than the red. Unless people stop hunting them for their furs, the colobi may all become extinct. They are found only in Africa.

5 / Mandrill   (*Mandrillus sphinx*)
These West African monkeys are immensely powerful and often quite aggressive. In zoos, where they are often seen, they can be dangerous. Their canine teeth are longer than those of a leopard. Mandrills can live for forty years if well cared for.

6 / Allen's Galago   (*Galago alleni*)
These small lorises from Africa are often called bush-babies. They are night creatures as you can tell from the size of their eyes. They have suction cups on the tips of their fingers and can leap surprising distances.

7 / Ring-tailed Lemur   (*Lepilemur catta*)
The lemurs, all of which come from the island of Madagascar off the east coast of Africa, are related to the monkeys. Like many other animals, they are threatened in their homeland because their forests are being cut down.

1 / Chimpanzee (*Pan troglodytes*)
If there is one animal in the zoo that is favored by the public above all others, it is the chimpanzee—nature's most accomplished clown. These African apes make terrible pets because although they are sweet and affectionate as babies (that is when you see them on television), they become quite unpredictable when they mature. Chimpanzees breed well in captivity and usually make excellent mothers. They are true apes, like all the animals pictured on the opposite page, and not monkeys.

2 / Lar Gibbon (*Hylobates lar*)
These small Asian apes, the smallest of the true apes, are extremely active forest animals. With their very long arms and small, slender bodies, the gibbons get around by brachiation—swinging hand over hand through the trees. The leaps they can take and the speed at which they can travel are quite amazing. These animals can do quite well in zoos, though they require expert care. Gibbons should have a diet that includes meat, fruit, green vegetables, and some form of grain and should have a cage large enough to allow them plenty of exercise.

3 / Orangutan (*Pongo pygmaeus*)
This magnificent great ape from Sumatra and Borneo is endangered and is no longer imported by zoos. The young orangutans seen in zoos today were born in this country. This animal, one of the most intelligent and spectacular in the world, may become extinct unless strong measures are taken to protect it in the wild. At the zoo, where it should have large quarters, the orangutan is a favorite exhibit, especially when it is a mother caring for her bright-eyed baby.

4 / Mountain Gorilla (*Gorilla gorilla*)
The gorilla is the largest of the apes and one of the most powerful animals in the zoo. Fortunately, zoo experts have learned how to care for these very special animals, and breeding in captivity is no longer uncommon. Some captive gorilla mothers refuse to take proper care of their delicate babies, and human nurses have to adopt them. Most captive gorillas are tricky to handle and can prove to be dangerous unless proper precautions are taken. Yet keepers become very attached to their gorilla charges, probably because they are so intelligent.

1 / Three-toed Sloth   (*Bradypus tridactylus*)
The tree sloths are a remarkable group of animals. They cannot walk on the ground, for they have been completely turned around and have legs designed to support them as they hang from trees. Sloths can crawl on the ground, but they aren't happy about doing it. Their sight and hearing are poor, and they depend on their senses of touch and smell to help them get around in the trees and find food—young leaves, buds, and tender twigs. These animals, who are believed to live about twelve years, are sometimes seen in zoos but can be difficult to keep. This species comes from South America.

2 / Giant Anteater   (*Myrmecophaga tridactyla*)
This is another strange South American animal, one that lives entirely on insects, termites, ants, and beetle larvae. Anteaters are large animals and will rear up on their hind legs if challenged and attempt to slash the enemy with their extremely powerful front claws. The claws, meant for tearing anthills apart, are also superb weapons. Anteaters are powerful swimmers and can move over open ground with considerable speed. When they sleep, they cover their heads and bodies with their great fanlike tails.

3 / Nine-banded Armadillo   (*Dasypus novemcinctus*)
A nine-banded is an American armadillo, a snuffling insect-eater that does well in zoos. Armadillos are fed a mixture of milk, eggs, and meat with vitamins added. Like the other animals pictured on the opposite page, they belong to a group of mammals known as the edentates. The scientific name Edentata (the order to which they belong) means "without teeth." That isn't accurate, though, since only the anteaters actually have no teeth at all. They are a strange and interesting group of animals that shows the extent to which mammals can diversify to take advantage of the world and the opportunities it offers animals that adapt.

1 / Kangaroo Rat   (*Dipodomys* sp.)
There are about two dozen species of kangaroo rats in the United States and Mexico. They are dry land animals and usually move about only at night. The direct sunlight would kill them if they tried to survive the desert by day.

2 / European Rabbit   (*Oryctolagus cuniculus*)
For some reason it is nearly impossible to get people to stop calling rabbits rodents. They aren't; they are lagomorphs. What is a lagomorph? It is a mammal similar to a rodent but with hind legs suited to jumping, such as a rabbit or a hare. The rodent category comprises rats, squirrels, and mice. This species of rabbit is found in Europe and parts of Africa.

3 / Chinchilla   (*Chinchilla laniger*)
This rodent originally came from South America, from mountainous areas of Chile and Bolivia. Chinchillas are now quite rare in the wild but are raised on farms for their lovely soft coats. They are clean and gentle little animals.

4 / Capybara   (*Hydrochoerus hydrochaeis*)
The capybara is the largest living rodent. A big specimen can weigh 110 pounds. It comes from Central and South America and looks like a gigantic guinea pig. In the wild it lives in bands of up to twenty animals. The capybara does breed in captivity.

5 / Prairie Dog   (*Cynomys ludovicianus*)
The name of this animal is prairie dog, and it does bark or at least yap, but there is no dog in it. It is a kind of squirrel and therefore a seed-eating rodent, not a carnivore. Prairie dogs live in "towns" or colonies in the wild, and that is the way they should be seen in zoos. They burrow deep into the earth and hibernate in the winter.

6 / Porcupine   (*Erethizon dorsatum*)
There are porcupines in Asia, Africa, and South America, but this species comes from North America. They are good exhibit animals, and they do not throw their quills the way the legends say. They are quiet zoo citizens and can be bred in captivity.

1 / Killer Whale   (*Orcinus orca*)
This splendid animal is really the largest of all the porpoises. A bull can be thirty-one feet long. Whales are extremely intelligent and can be trained to do a variety of stunts. Some large aquariums have been keeping them but usually without too much luck. They need a very large body of water and eat many, many pounds of fish a day. The stories about their being man-eaters are apparently just so much nonsense.

2 / Beluga   (*Delphinapterus* sp.)
Some aquariums and zoos have been exhibiting these beautiful white whales, and in some cases the animals have survived for a number of years. The belugas come from far northern waters and often travel in large schools or pods. Like all whales and dolphins, they are intelligent, responsive to both good care and training, and always popular with the public. *Never* throw anything into the water in a whale's enclosure. The simplest piece of scrap may be gobbled up and can prove fatal. There have been such cases.

3 / Common Porpoise or Dolphin   (*Delphinus delphis*)
This intelligent and gentle animal is one of the species sometimes seen in porpoise shows in aquariums and very large zoos. These animals learn very quickly, almost instantly once they are shown what is expected of them, but they seem to find it impossible to "unlearn" something. They get very upset if you try to get them to forget an old way of doing things and try a new. They are sleek, fast, and handsome animals and should have a large tank in which to exercise. They must remain in the water at all times, of course, for although they are air breathers, it is bad for them if their skins get dry.

1 / Striped Hyena (*Hyaena hyaena*)
The striped hyena is widespread and quite common in some isolated areas of Asia and Asia Minor. Many people have the wrong idea about these animals. They think they are only scavengers, but the hyenas are powerful and active hunters and have the strongest jaws of any hunting land animal. They can crack the largest bones.

2 / Spotted Hyena (*Crocuta crocuta*)
This is the laughing hyena whose weird call in the middle of the night has made many an African traveler's hair stand on end. Like their striped cousins, they are hunters. A big male can weigh eighty-five pounds. They settle down and do well in zoos and can be bred in captivity.

3 / Gray Wolf (*Canis lupus*)
This proud animal has a very unfair reputation. Gray wolves are excellent parents, hardly ever kill anything they are not going to eat, never attack man, and very seldom fight among themselves. They are now rare south of Canada, which doesn't speak well for the wisdom of man. They breed in zoos, and the young are very interesting to watch as they grow up and learn to get along in their parents' pack.

4 / Fennec (*Fennecus zerda*)
This is an African and Asia Minor species and a delightful little animal to see in zoos. Its ears seem too large to be real, and it is alert and interested in everything that goes on around it. Some zoos have succeeded in breeding fennecs.

5 / Gray Fox (*Urocyon cinereoargenteus*)
Gray foxes are interesting animals. They can climb trees almost as well as cats can. They are active and intelligent and settle down well in the world of the zoo. They come from North America, but are quite different in appearance from our red fox. The gray is the smaller of the two.

1 / Grizzly Bear   (*Ursus arctos horribilis*)
This great bear is now rare in the original forty-eight states. It is often seen in zoos, though, and is regularly bred in captivity. Grizzly bears are one of the most powerful land animals and must be watched very carefully by the keepers who work with them. They like water, and they like to eat. They do not really hibernate but do sleep most of the winter.

2 / Polar Bear   (*Thalarctos maritimus*)
The polar bears always attract a crowd at the zoo. They are found all around the top of the world and are becoming scarce because of men who hunt them for their magnificent white coats. In zoos they should have a very large pool to swim in. As a matter of fact, the polar bears' enclosure should be 90 percent water and 10 percent land. They cannot be trusted.

3 / Giant Panda   (*Ailuropoda melanoleuca*)
These splendid animals are found wild only in China and have always attracted more attention than any other single zoo species. They have been bred in Chinese zoos but nowhere else, and only one zoo in America has pandas now. No one knows whether they are really rare or not, but we should assume they are. They are not bears, but sort of halfway between the bear and the raccoon family.

4 / Raccoon   (*Procyon lotor*)
Every zoo seems to have a colony of these interesting little busybodies because they are one of the most common of American wild animals and make excellent zoo exhibits. They settle down, do very well, and produce lots of young. They belong to a family of meat-eating mammals known as the Procyonidae.

5 / Sun Bear   (*Helarctos malayanus*)
The sun bear comes to us from Burma, Indochina, Thailand, the Malay Peninsula, Sumatra, and Borneo. That is a wide area, and this animal, the smallest of the world's bears, enjoys great popularity in the zoo. Some zoos have succeeded in breeding them!

6 / Black Bear   (*Eurarctos americanus*)
This is the common North American bear and one of the most common bears in our zoos. Black bears breed well, live long, and make interesting exhibits. We call them "black" bears, but they can also be cinnamon, blue, and even white—depending on where they come from. Next to our deer, this is America's most common large animal.

1 / African Linsang (*Poiana richardsoni*)
This West African meat-eater is a relative of the mongoose. It spends a great deal of time in trees and makes nests to sleep in, but the nest is used for only a few days and then the animal moves on to build another. It is a forest animal and is active only at night. If you lived in West Africa, you would probably never see one. In the zoo you at least have a chance!

2 / Spotted Skunk (*Spilogale putorius*)
The skunks are members of the weasel family, but they have been equipped with a special weapon of defense. In the zoo a simple little operation makes the skunk safer and more pleasant to live with.

3 / River Otter (*Lutra canadensis*)
This large and very active weasel is a clown, a show-off, and a tease. As a result, otters are fun to watch. They swim like torpedoes and dive like dolphins. They are a favorite zoo exhibit and require plenty of clean, cool water. This species comes from North America, but there are South American, Asian, and African species, too. A number of kinds are seen in zoos.

4 / Weasel (*Mustela* sp.)
There are perhaps a dozen different kinds of weasels exhibited in zoos, and this picture is of a "generalized" weasel representing them all. They are active, nervous, and not terribly pleasant animals with enormous appetites and quick tempers. They breed well, though, and are sleek and attractive. The mink is a weasel, for example.

5 / Civet (*Viverra* sp.)
The civets are also relatives of the mongoose belonging to the family Viverridae. They come from many parts of Asia and are usually active only at night. People sometimes call them civet cats, but they are not cats at all. Just look at the drawing.

6 / African Mongoose (*Herpestes ichneumon*)
There are many kinds of mongooses (not mongeese!), and this species from Africa is typical. They are active animals, and they do battle cobras—simply because they like to eat snakes as well as birds and small mammals. Mongooses have big appetites and spend a good part of their time hunting. They are often kept as house pets to keep snakes and rats under control.

1 / Leopard (*Panthera pardus*)
This magnificent cat is found in both Asia and Africa and is very adaptable. In zoos it can be a problem because it is so athletic. A leopard can leap over moats and fences that will easily hold another cat. That is why it is almost always seen in cages even when the other cats are in open areas. The black leopard or panther, by the way, is the same animal as the spotted leopard, just a darker phase. Look at a black leopard in the slanting sunlight and you will see his spots.

2 / Jaguar (*Panthera onca*)
The jaguar is the largest cat in the Western Hemisphere. It is a powerful hunter from Central and South America. On rare occasions one ranges as far north as Texas or Arizona. The jaguar, like the leopard, is found in a melano form—seemingly black. The jaguar, too, is an athlete and is usually confined to cages or enclosures with a closed-in top.

3 / Lion (*Panthera leo*)
The lion is often called the king of the beasts, but it isn't a king at all! A lion will give way to an elephant, a rhino, and often to a Cape buffalo. Lions are also referred to as jungle animals, but they are never found in the jungle! They are open plains cats and stay out of the deep forest. They can climb trees quite well, although many people don't realize that. All in all, the lion is an animal full of surprises.

VANISHING ANIMAL

4 / Tiger (*Panthera tigris*)
People think of tigers as tropical animals, but in fact they have arrived in tropical lands like India only recently as evolutionary time goes. They started in the snows of Siberia and northern China. The largest of all living cats, the tiger really doesn't like the heat. On very hot days, jungle tigers sit in a river to cool off.

5 / Puma (*Felis concolor*)
This lovely cat of the New World has the greatest north-south range of any wild cat—from Canada to southern South America. It is the same animal that people call panther, painter, cougar, mountain lion, and catamount. It is one of the most athletic of all cats, and although very well behaved as a citizen of the zoo, the puma is usually found in a very well-enclosed area.

1 / Bobcat (*Lynx rufus*)
This small North American wildcat is a popular member of the zoo family. It settles down in captivity, and a number of zoos have managed to breed it. Because it is so clever and so adaptable, the bobcat is one species that doesn't seem in danger in the wild. It seems able to live almost anywhere and can get by on wildlife and the occasional domestic animals it manages to steal. It is a smart-looking and interesting animal.

2 / Cheetah (*Acinonyx jubatus*)
This cat is unlike any other cat in the world. It has legs like a dog and claws that cannot be retracted into sheaths. It is the fastest land animal in the world and can probably run sixty miles an hour for a short distance. It has very rarely been bred in captivity and is causing scientists some concern. It could become extinct. It is already gone from almost all of Asia and much of Africa. It would be a tragedy if this animal should vanish from our planet. It adapts well to captivity—now if only it would breed!

3 / Canadian Lynx (*Lynx canadensis*)
Along about the Canadian border the bobcat starts becoming rare and the larger Canadian lynx takes over. This is the second largest cat found regularly in North America north of Mexico, the largest, of course, being the puma. The lynx has large, broad feet, and it gets along in deep snow and freezing cold. It isn't seen around farms and towns as much as bobcats are, as it is shier and more secretive. The lynx doesn't like hot climates, but it does well in zoos, particularly in the north.

1 / California Sea Lion   (*Zalophus californianus*)
This is the trained "seal" of circuses and aquariums. It is a species commonly seen in captivity and is very intelligent. The sea lion, along with the seals and walruses, belongs to a specialized group of mammals known as the pinnipeds—the fin-footed ones. Their diet, of course, consists of fish.

2 / Alaskan Fur Seal   (*Callorhinus ursinus cynocephalus*)
This is one of the very large seals, and it is hunted for its valuable fur. Hunted isn't such a good word since all that is involved is walking up to these animals on the beach and killing them. They are seen in only the largest zoos and don't do too well in captivity in most situations. There have been some successful cases, however, and one American zoo claims to have bred them.

3 / Walrus   (*Odobenus rosmarus*)
This absolutely enormous member of the seal family can weigh almost a ton and a half. As young, walruses are very sweet and become attached to their keepers. They can be tricky to handle when they get older since they don't always know their own strength and weight. Many zoos and aquariums have tried to keep them, and some have succeeded very well. They must have their own large pools, however, and they can be killed by people throwing things into the water. One died because it swallowed a child's rubber ball. *Never* throw anything into an animal enclosure.

4 / Manatee   (*Trichechus manatus*)
Some zoos and aquariums manage to keep manatees, but they require a large pool all to themselves. They must be in the water at all times, though they are mammals. Several factors are causing manatees to become scarce. People have hunted them, their water gets polluted and they must move on or die, and they have accidents with motorboats. They are from the New World but are related to an Old World species called dugongs, which are scarce now, too.

1 / African Elephant   (*Loxodonta africana*)
The bull African elephant is the largest land animal left on this planet. A big specimen can weigh over seven tons. The African elephants have much heavier tusks than the Asian species and much larger ears. Their foreheads slope, and they don't have the prominent bumps the Asian elephants have. African elephants are not very often seen in zoos and full-grown bulls almost never because they would be much, much more than even the best zoo would want to care for in most instances. Their strength cannot even be estimated, and controlling such an animal could be almost impossible. There is, however, no more impressive animal in the world.

2 / Rock Hyrax   (*Heterohyrax syriacus*)
This nine- to ten-pound animal, believe it or not, is the elephant's closest living relative, although the two aren't really very close. The rock hyraxes have funny little tusks, usually hidden by their upper lips, and elephant-like feet, yet it is difficult to think of them as cousins to the elephant. In the Bible these animals are called coneys. They are found in Asia Minor and Africa and feed mainly on roots and bulbs. They are occasionally seen in zoos.

3 / Young Indian Elephant   (*Elephas maximus*)
This is the species generally seen in zoos. They are rarely bred in captivity, but there have been exceptions. The babies you see, though, were almost certainly brought from their native country—any one of many in Asia including India, Bangladesh, and Ceylon. They make excellent zoo exhibits, and although one will occasionally become difficult, Asian elephants are usually easy to handle—if you are an expert. They live to be sixty years old and never fail to please their many admirers. Their quiet nature shouldn't fool you, however. Any keeper can tell you that they are extremely powerful. A single captive elephant may eat as much as four hundred pounds of food a day—and consume fifty gallons of water.

1 / Malayan Tapir   (*Tapirus indicus*)
This very interesting animal is a remote relative of the horse and is found in Asia. The other three kinds of tapirs live in South and Central America. All tapirs are becoming rare, but fortunately this species is now being bred in a number of zoos. This is one form of wildlife that will probably survive only in captivity in the years ahead. The babies look nothing like this adult. They have stripes and spots on a dark background that help them hide in thick vegetation.

2 / Baird's Tapir   (*Tapirus terrestris*)
This form of tapir is found in Mexico, in Central America, and down into South America. It is becoming rare all over its range because it is being hunted and because its habitat is being destroyed. It can be bred in captivity, although that is a rare event. Like all tapirs, this one could vanish from the wild in our time. It is a vegetarian and very shy in the wild. In captivity, the animal is quite calm, but it can and does bite!

3 / Przewalski's Horse   (*Equus przewalskii*)
The Przewalski or the Mongolian wild horse is the only truly wild horse left in the world. The horses we think of as wild—the mustangs—are feral, descendants of domestic horses that got away. The Przewalski, though, is truly wild and extremely rare, and in the years ahead the zoo will be the only place where it will be able to survive.

4 / Grant's Zebra   (*Equus burchelli böhmi*)
This is the most common zoo zebra and is considerably smaller than the Grevy's zebra also shown on the opposite page. The Grant's will seldom weigh more than seven hundred pounds (unless it is being overfed), while a Grevy's can weigh close to a thousand pounds. In this zebra the stripes are broad and very clearly marked.

5 / Grevy's Zebra   (*Equus greyvi*)
This large, handsome zebra looks something like a mule—with stripes, of course. Like all zebras it is found only in Africa. Its stripes are much narrower than those on Grant's zebra, and that characteristic, along with its large size, makes it easy to distinguish in the zoo—and in the wild. The babies have stripes, too, but they are brown and white instead of black and white.

1 / Black Rhinoceros   (*Diceros bicornis*)
The black rhinoceros isn't black any more than the white rhinoceros seen also on the opposite page is white. They are both gray! The so-called black rhino, a large animal that can weigh over two tons, is found in parts of eastern and southern Africa. It is aggressive at times, although stories on this subject are generally exaggerated. Still, keepers have to be on their guard and get to know their charges very, very well. There are fewer of these great animals in Africa every year, and unless care is taken, we may see the last of them in our time. They do well in zoos, however, and even breed in captivity.

2 / Indian Rhinoceros   (*Rhinoceros unicornis*)
The proper common name for this marvelous animal is the Great Indian One-horned Rhinoceros, and it deserves the name *great*. It is one of the largest living land animals and unfortunately is becoming more and more rare. Few zoos are able to obtain specimens, and if you are fortunate enough to visit a zoo that has some Indian rhinos, look at them well. It is most unlikely that your grandchildren will be able to do so. Compare it with an African rhinoceros and see how different the skin folds are.

3 / White Rhinoceros   (*Ceratotherium simum*)
People insist on calling this animal the white rhinoceros, although that really isn't the proper name. Either square-lip rhino or broad-lipped rhino would be better. This gigantic animal (up to 6,500 pounds in a large bull) comes from South Africa and was once seriously endangered. It is now protected and is breeding well in game parks. Surplus animals are shipped to zoos all over the world, so that you may encounter a pair in a zoo that you visit. Few animals are more impressive than this one. Interestingly enough, the square-lip is usually quieter and more sensible in captivity than the black rhinoceros—usually, but not always.

1 / Collared Peccary (*Tayassu tajacu*)
The collared peccary is a powerful wild pig, the only one native to our shores. The wild boar we hear so much about is an import. The peccary is a dry-area animal that will eat almost anything, including rattlesnakes. Peccaries live very close family lives, and the adults will attack with great determination if they or their young are threatened. They make interesting zoo animals and can settle down and become fairly tame.

2 / Hippopotamus (*Hippopotamus amphibius*)
The name means "river horse," but this huge African animal is hardly a horse! It can weigh almost five tons, making it one of the heaviest animals on earth. Hippopotamuses do well in zoos but require large enclosures with plenty of water. They can be injured by strong sunlight when they don't have the opportunity to submerge and avoid the most dangerous rays. Hippos are quiet, generally sensible zoo citizens, although they can be quite dangerous in the wild.

VANISHING ANIMAL

3 / Pygmy Hippopotamus (*Choeropsis liberiensis*)
This damp forest animal is found in western Africa and is not nearly so big as the other hippo and not as well known. The pygmy hippo doesn't usually weigh more than six hundred pounds, although zoo specimens can get a little overweight from inactivity. A number of zoos have been able to breed pygmies in captivity, and that is always a happy event. They are getting scarce in the wild where some natives still hunt them for food.

1 / Llama (*Lama peruana*)
One of the most popular of zoo animals, South America's mountain version of the camel comes in many colors and markings. The llama isn't always very pleasant, but quiet specimens can be placed in zoo "contact" or petting areas. There are no wild llamas left in the world, except some that may have escaped. By the time the Spanish reached South America, all llamas had been domesticated.

2 / Dromedary (*Camelus dromedarius*)
This is the one-humped camel whose exact range will never be known. There are no wild dromedaries left in the world—all are under domestication. Dromedaries probably originated in Arabia, but we don't know that for sure. This is the famous "ship-of-the-desert" camel and a favorite zoo citizen. (They don't have very nice personalities. They spit.)

3 / Bactrian Camel (*Camelus bactrianus*)
This is the two-humped camel, which came originally from Turkistan and Mongolia. These camels have shorter legs than the one-humped dromedary, but it is the design of their backs that distinguishes them so easily when we see them in the zoo. They have been domesticated for a very, very long time. It may be that all bactrian camels now running wild (there aren't many in Mongolia any more) are descended from camels that were once owned by man. We would call them *feral* in that case.

4 / Guanaco (*Lama guanacoe*)
The guanaco is one of the two relatives of the camel still found wild in South America. The vicuna, also found in South America, is the other, and it is getting to be very rare. Small herds of wild guanaco can still be found running wild in areas as high as five thousand feet above sea level. There are no relatives of the camel still found in North America. They became extinct long ago.

1 / Elk (*Cervus canadensis*)
Now this is where common names can get confusing. What we call an elk is really a wapiti—that is pronounced wha-*pee*-tee. What we call a moose is what people in Europe call an elk. They don't use the word *moose* at all. That is why scientific names are needed. Anyway, our elk or wapiti is a splendid member of the deer family, meaning it has antlers, not horns. Those magnificent racks are dropped every year and grown again in time for the mating season in the late fall.

2 / Moose (*Alces alces*)
We call it moose, the Europeans call it elk; either way it is the largest and most impressive deer in the world. It is found in northern North America and in some northern areas in Europe. Not too many zoos have had luck with moose. They don't generally do very well and die shortly after arrival. Some zoos have been successful, however, and some have even bred them. This is a splendid animal, although the bulls can be quite dangerous in the late fall when it is time to mate.

3 / Woodland Caribou (*Rangifer tarandus*)
Here is another place where confusion is common. The caribou and the reindeer are the same species of animal. The caribou are found in the Western Hemisphere and in Siberia. The reindeer are from Greenland and Europe. For some reason the reindeer are easier to domesticate. These interesting and important animals are often seen in zoos where they can do quite well, though they are better off in colder climates.

4 / Axis Deer (*Axis axis*)
The lovely spotted axis deer come from India and Ceylon, and many people think they are the most beautiful of all deer. They keep their spots all through their lives. Also called *chital*, they are regularly bred in zoos all around the world, where they are a common and popular exhibit. They are commonly seen in herds like many other kinds of deer.

5 / Water Chevrotain (*Hyemoschus aquaticus*)
Just over a foot tall at the shoulder, this little hoofed animal is a very rare zoo exhibit. You will be lucky to see a pair of these small African animals. As their name implies, they swim very well and are found near jungle rivers. They dive in and under the water to escape danger. Compare this hoofed animal with the moose also pictured on the opposite page.

1 / Reticulated Giraffe   (*Giraffa camelopardalis*)
The giraffes belong to the order Artiodactyla, the even-toed hoofed animals. They are, of course, the tallest animals in the world and are found only in Africa. A big bull can weigh almost two tons and has a kick that can kill any predator that tries to attack him. Despite their extremely long necks, all giraffes have the same number of neck bones as you do—seven. Those bones are rather large, however. Giraffes are strictly vegetarians and can eat the thorniest trees without hurting themselves. They feed on the parts of the trees no other hoofed animals can reach. They are above competition.

2 / Okapi   (*Okapia johnstoni*)
This dense forest animal from Africa is the closest relative the giraffe has. It was one of the last of the large animals to become known to science. Before 1900 they were thought to be a tall tale made up by explorers trying to attract attention to themselves. But then the animal itself was discovered much to the surprise of the entire scientific world. They are handsome creatures and are seen in some of the larger zoos.

3 / Pronghorn   (*Antilocapra americana*)
This handsome western American plains animal is often called an antelope, although it really isn't an antelope at all. (There are no antelopes in the Western Hemisphere.) It is the fastest land mammal in the Americas and can run at forty miles an hour. It is a strange animal because it is the only one that sheds its horns. Other animals shed antlers—never horns. Pronghorns are difficult to keep in zoos because they panic easily and can kill themselves by running into fences, walls, and trees. Some zoos have had success with them, however.

1 / Wild Yak   (*Bos grunniens*)

The yak or grunting ox comes from China—from an area called Kansu. It is also found in Tibet and some parts of India. It is now rare and found only in some of the larger zoos. There are domestic yak, of course, and they are seen in zoos fairly often. A true wild yak bull may be twice as large as a domestic yak. That makes quite a difference.

2 / American Bison   (*Bison bison*)

There is probably no way to get people to stop calling our bison a buffalo. Actually, there are no native buffalo in North America. Buffaloes come from Asia and Africa—and they, too, are pictured on the opposite page. Our bison once numbered close to 100 million, but by the beginning of this century they faced extinction. Controlled captive breeding brought them back, and there are probably 35,000 to 40,000 in America today. Some are slaughtered for meat each year. There are lots of them in zoos, and they make a fine exhibit since they were so important in American history. They are not very pleasant animals, and the keepers have to be wary of the bulls. A big bull can weigh a ton and a half.

3 / Water Buffalo   (*Bubalus bubalis*)

A great many people think the water buffalo comes from Africa, but that is wrong. The Cape buffalo is the African species; the water buffalo is from Asia. There are wild buffalo and domestic buffalo, and they are used for almost everything. They are slaughtered for their meat and hides and used to pull carts and plows. Children even ride them to school! The water buffalo you see in your zoo may be either a domestic variety or a wild one. It will be hard for you to tell. Either way, the water buffalo is a large and interesting animal that is very important in the lives of people in Asia. In many areas, people would die without them.

4 / Cape Buffalo   (*Syncerus caffer*)

This is one of the most impressive animals in Africa and one of the most impressive in the zoo. Cape buffalo are bred in zoos from time to time, and the inquisitive little "buff" with his oversize ears gives little hint of what he will one day be—a very large, extremely powerful, and sometimes tricky animal.

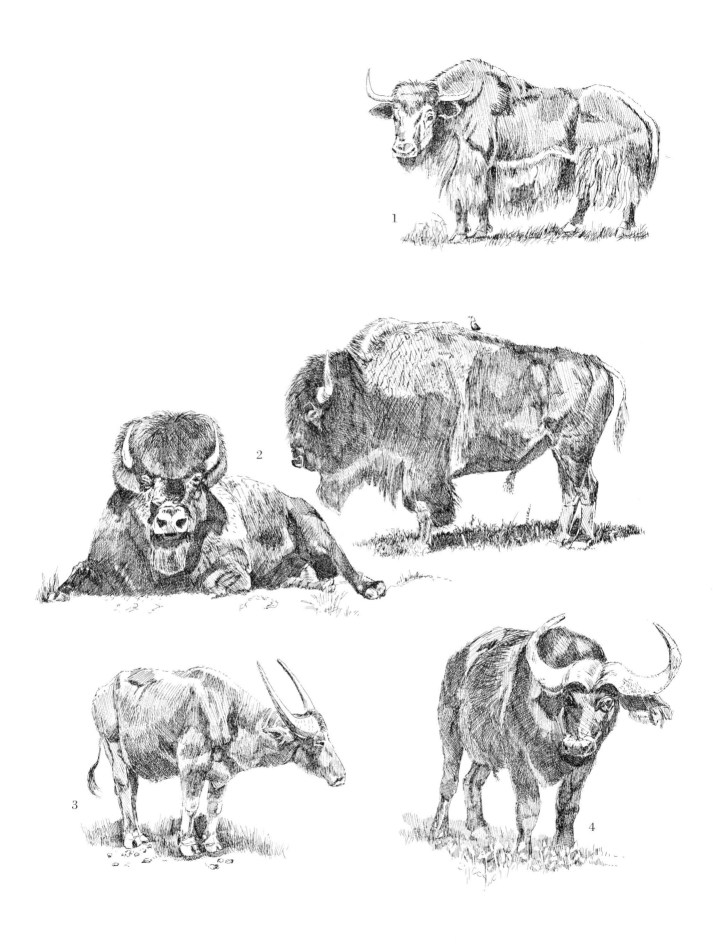

1 / Sable Antelope   (*Hippotragus niger*)
Sable antelopes are among the most impressive antelopes found in Africa—and certainly in zoos. They are beautiful animals and can weigh over 650 pounds. Their magnificent horns sweep up and back and are very potent weapons for use against an attacker. Fortunately, sables breed well in captivity. They can run thirty-six miles an hour when frightened and require large zoo enclosures. In the wild they are often seen in herds of ten to twenty animals.

2 / Black Buck   (*Antilope cervicapra*)
This is an open plains antelope from India and Pakistan. It is beautifully marked, among the most beautiful of all hoofed species. Happily, it breeds well in captivity because it is now quite rare in the wild. Most antelopes do not have a great color variation between males and females. In this species there is a difference, though, for the male is much darker. Only the males have horns, which can be over two feet long.

3 / Giant Eland   (*Taurotragus oryx*)
These great African animals prefer gently rolling countryside with only scattered bushes and trees. They are very large—a bull can weigh almost a ton. In Africa they travel in herds, and as many as a hundred or more may be seen at one time. There is a plan now to domesticate these animals and use them for beef in areas where cattle do not do well. Elands are much hardier than cattle. They make a fine zoo exhibit and often bear young year after year.

4 / Brindled Gnu   (*Connochaetes taurinus*)
This is a very disagreeable animal, and its proper common name is wildebeest. It is found in East Africa in absolutely enormous herds—tens of thousands during the migrations across the Serengeti Plains. These animals are a favorite food of the lion, the hyena, the Cape hunting dog, and the leopard, but there always seem to be enough gnus to go around. It is dangerous for a keeper to go into an enclosure with this animal. It charges on the slightest provocation and can use its sharp horns very effectively.

1 / Dorcas Gazelle (*Gazella dorcas*)
These beautiful little animals travel in herds and cover open ground at great speed when startled. They are shy animals but can do well in a really good zoo that has room for them. They shouldn't be kept in very small pens. Dorcas gazelles are found in the Holy Land and are referred to in the Bible. In Israel they are often seen in pastures with cattle. The farmers don't seem to mind.

2 / Thomson Gazelle (*Gazella thomsoni*)
The "tommie" is one of Africa's most delightful hoofed animals. It seldom weighs much over 55 pounds, but it is hardy and fast. Tommies are a favorite food of the lion and the other African predators. They are bred in zoos, and baby tommies are among the most attractive and popular exhibits. They are smart little animals, and when they have something serious on their minds, they swish their tails rapidly from side to side. It is probably a signal to other members of the herd to keep alert.

3 / Gerenuk (*Litocranius walleri*)
The gerenuk (or giraffe gazelle, as it is sometimes known) is a very large gazelle and can weigh as much as 115 pounds. It has an exceptionally long neck and the ability to stand on its hind legs, which allows it to escape competition and feed on the tops of bushes where the other animals cannot reach. Gerenuks are seen in only the largest zoos and are a very special exhibit. They have been bred in zoos from time to time. (The illustration of the head shows the male of the species.)

1 / Aoudad   (*Ammotragus lervia*)
These splendid wild animals are also known as the Barbary sheep, and they are commonly seen in zoos. They make a fine, interesting exhibit, but the animals should always have in their enclosure a small "mountain" that they can climb and scramble over. Their ability to move along narrow cliff edges is quite remarkable. They are native to North Africa from the Atlantic Ocean all the way to the Middle East. This is the only wild sheep native to Africa.

2 / Rocky Mountain Goat   (*Oreamnos americanus*)
Only about ten zoos in the world now exhibit this splendid North American wild goat. When you see one, you are looking at one of the finest animals in the Western Hemisphere. Its speed and agility on narrow mountain ledges is legendary. Its feet have special pads that enable it to jump down from heights that would kill or at least disable other animals.

3 / Rocky Mountain Sheep   (*Ovis canadensis*)
This is one of the bighorn sheep, related to animals found in Russia and Asia. Rocky Mountain sheep are seen in some zoos and should have a rock pile to climb. They are among the most cherished of all big-game trophies, although they are certainly far more beautiful alive than dead. The rams fight at breeding time and can do serious damage. Usually, though, the inferior male gives way and allows the victor to have all the females to himself.

4 / Musk Ox   (*Ovibos moschatus*)
This is an Arctic animal that lives in both Canada and Greenland. It isn't really very closely related to any other animal, and its history is something of a mystery to scientists. Musk oxen travel in herds, and when danger threatens, they gather into a tight circle with the males on the outside, shoulder-to-shoulder, with their horns facing outward. They are short-tempered animals, but some zoos do fairly well with them. They are bred in captivity and may one day be used as a beef animal in the Far North.

1 / Emu   (*Dromiceius novaehollandiae*)
This large bird was once found all over Australia. It is the second largest of the ratites or large flightless birds and can stand six feet tall. It can weigh 120 pounds in what would be an exceptionally large specimen. The hen may lay as many as sixteen huge greenish eggs at a time, althought eight to ten would be closer to average. Emus are interesting zoo animals.

2 / Common Rhea   (*Rhea americana*)
What the emu is to Australia and the ostrich is to Africa, the rhea is to South America. A large rhea may stand five feet tall and weigh as much as fifty-five pounds. It is an extremely rapid runner and when frightened will dash off with its head almost straight out in front. It would take a man on a horse to keep up with it. The rhea cannot, of course, fly.

3 / Cassowary   (*Casuarius* sp.)
The cassowaries are Australia's other great running bird. But, unlike the emu, they are also found in New Guinea and a few other islands in the area. They have very powerful feet, and the middle toe on each foot has a long, sharp claw. It can be a deadly weapon, and keepers are well aware of it when they have to enter an enclosure with these birds, which can weigh over a hundred pounds.

4 / Emperor Penguin   (*Aptenodytes forsteri*)
The emperor penguin is the world's largest and heaviest sea bird. Full-grown specimens can stand three feet tall and weigh a hundred pounds. Penguins are always popular in zoos, although many species, like the emperor, can be hard to maintain. They are sensitive to disease and fungus infection. They don't have such problems in their native habitat, which is the Antarctic region of the Southern Hemisphere. Not many people realize there are none anywhere near the North Pole.

5 / Ostrich   (*Struthio camelus*)
The ostrich, the largest bird in the world, comes from Africa—from open plains areas and never the forests or jungles. An adult can stand eight feet tall and weigh 345 pounds. That is a true giant, and a kick from one of these birds (they kick forward like a football player) can put a man in the hospital. Zoo keepers bear that in mind! (Males are black and white, females grayish brown.)

1 / Rufescent Tinamou   (*Crypturellus cinnamomeus*)
The tinamous—there are about fifty kinds—come from Mexico and Central and South America. Their eggs are among the most beautiful in the world. They glisten and are usually solid in color—chocolate, purple, or dark bluish green. The brown and gray birds themselves are not as attractive as their eggs seem to promise. If you get to know the keeper in the bird house, he may show you a tinamou egg—they often have some around.

2 / Arctic Loon   (*Gavia arctica*)
This bird is also called a diver. And don't let the word Arctic fool you. It lives there only in the summer. When it gets really cold up north, you can find Arctic loons in Mexico! The glossy black and white, long-necked loon has an incredible ghostlike voice, a combination of moan, whistle, and crazy laugh. But you won't hear it in the zoo—just on a foggy lake at night where it can really scare the wits out of you.

3 / Great Crested Grebe   (*Podiceps cristatus*)
The grebes are found all around the world (this one comes from the Old World), and although they are large, handsome water birds, they are neither ducks nor geese. This is one of the largest grebes, about eighteen inches long. It has a shocking pink bill and a very long neck.

4 / Brown Pelican   (*Pelecanus occidentalis*)
Unfortunately, when you see the pelican in the zoo, you will not see him at his best. Clumsy and even clownish on the ground, a pelican in flight is beautiful and graceful. However, it is good to see a bird as fascinating as this close up. This species, the brown pelican, once common in Florida, is in danger of extinction because of the pollution we have spread around, and it would be terrible if our DDT and other chemical rubbish killed the pelicans off. The next time you see one in the zoo, look at it carefully. It is a special kind of bird.

1 / Purple Heron   (*Ardea purpurea*)
The herons are splendid, handsome birds. There are 64 different kinds, and some of them can stand over five feet tall. Some species breed in the zoo. The purple heron shown on the opposite page comes from Europe and Africa and lives in marshes and swamps. It has a wingspread of almost four feet.

2 / Least Bittern   (*Ixobrychus exilis*)
This bird can stand in the reeds at the edge of a swamp and imitate the reeds exactly—even their movement. When the wind makes the reeds move, the bittern moves to the same rhythm, as the bird is shown doing on the opposite page. This small member of the heron family ranges over much of the United States. Its call is a rapid *coo-coo-coo*.

3 / Snakebird   (*Anhinga* sp.)
The snakebird (so called because of the shape of its neck) has another strange name—water turkey—and in some places it is called a darter. Without science, names can be puzzling. The label in the zoo may just read anhinga, though, for that really is a less confusing name. The anhingas are found around the world. They are powerful underwater swimmers and hunt fish, frogs, and other fresh-water animals.

4 / Hammerhead Stork   (*Scopus umbretta*)
This African bird isn't really a stork. It is always found around water, and its nest, built up in a tree, is gigantic. It is very exciting for zoo people when they can get birds like this to build nests in captivity. This dusky brown species is only about two feet long but has a thick crest that gives the bird its name.

5 / Shoebill or Whale-headed Stork   (*Balaeniceps rex*)
This enormous African water bird isn't exactly beautiful, but it certainly is interesting looking. It inhabits a number of areas in Africa and is mainly gray in appearance with a wide bill. Note the many different shapes and sizes the bills of storks take. Make sketches of them and then test yourself a week later and see if you remember which bill belongs to which kind of bird.

1 / White Stork   (*Ciconia ciconia*)
One way of telling the stork from the heron is by the position of the neck during flight—the heron's is drawn back while the stork's extends out in front. This stork breeds in Europe—often in chimneys—and winters in Africa. It is the same stork people used to say brought babies. It is black and white with a red bill and red legs.

2 / African Spoonbill   (*Platalea alba*)
There is a European spoonbill that looks like the African species shown on the opposite page. It is easy to tell them apart, however, because the African has red legs and the European has black legs. Both have white plumage. Those long legs, of course, enable the bird to wade through shallow water and have the good high view it needs when hunting fish and other small aquatic creatures.

3 / Flamingo   (*Phoenicopterus ruber*)
Flamingos are among the most attractive and popular zoo birds. In the wild they are very gregarious, and on some African lakes there may be two or three million at a time! In the zoo, when flamingos start looking pale, an extract of carrots is added to their diet to make their rose-pink color deepen. We have some wild flamingos in Florida. They are bright pink.

4 / Sacred Ibis   (*Threskiornis aethiopicus*)
The ibis isn't a very beautiful bird, but it is interesting looking, and it is said the ancient Egyptians worshiped this bird in the form of a god. It has a naked black head and neck and white plumage and is found in marshes, swamps, and pastures in a number of African areas.

1 / Mute Swan   (*Cygnus olor*)

This is a European swan that was brought here to decorate garden pools and private estates. These mute swans managed to get loose over the years, and now America has a good wild population of them. They are commonly seen in and out of zoos. The babies are called cygnets, and the male is called a cob. They are always near water, of course, and are snowy white all over with an orange bill.

2 / Black-necked Swan   (*Cygnus melanocoryphus*)

There are between six and eight different species of swans in the world—scientists have been arguing about the correct number for years—and the black-necked from southern South America is one of the most strikingly handsome. It is very popular in zoos and is easy to breed. Its black neck makes it easy to distinguish from all other swans.

VANISHING ANIMAL

3 / Nene   (*Branta sandvicensis*)

If it weren't for zoos, there wouldn't be any nenes left in the world. Also called Hawaiian geese, they were on the verge of extinction only a few years ago. They weigh four or five pounds and are found in the high lava fields in Hawaii. They aren't as tied to water as most other geese are.

4 / Canada Goose   (*Branta canadensis*)

This large goose from all across Canada often will see geese and ducks on a zoo pond and settle down to join them. Zoos very often have wild waterfowl join their collection. Many of them migrate, and the collection can vary from season to season. This is a large goose, gray, white, and black.

5 / Wood Duck   (*Aix sponsa*)

This exceptionally lovely bird is the most brightly colored of all North American ducks and one of the handsomest ducks in the world. It is iridescent with a white belly, dark breast and wings. There are no words to describe its jewel-like colors.

6 / Canvasback Duck   (*Aythya valisineria*)

The canvasback is a kind of diving duck who submerges in search of underwater plants and little animals. In this illustration, the male is on the right. He has a rusty-red head and neck and a black breast. The female, on the left, is grayish. In most species of birds the male is more colorful, to attract females at mating time.

1 / Harpy Eagle   (*Harpia harpyja*)
The birds of prey are among the most popular birds in the zoo. They seem to capture the imaginations of all who view them. The harpy eagle, a great hunting bird from Central and South America, is a very interesting exhibit. It is a somber gray-brown bird with a strange crest of feathers. Unhappily, birds of prey or raptors don't breed well in captivity.

2 / Bald Eagle   (*Haliaeetus leucocephalus*)
Everyone enjoys seeing our national emblem, our splendid bald eagle. Bald eagles are seen in many zoos, but breeding is a problem. They are large birds, up to thirty-four inches long. Only the adults have the white head and neck. They are scavengers and are especially fond of fish.

3 / Peregrine Falcon   (*Falco peregrinus*)
This bird is rarely seen in zoos. In fact, it is now rarely seen anywhere! This is the royal falcon, the fastest bird in the world. Once, only kings could own them. It is said that when they dive to attack ("stoop"), they can go 150 miles an hour. They are about the size of a crow and are slate-colored.

4 / Red-tailed Hawk   (*Buteo jamaicensis*)
This imposing North American hawk or buteo is a fine-looking bird, fierce, brave, and determined. The red part of the tail is on top, but the bird veers as it flies and the red can be seen from the ground. It is a large hawk, about two feet long.

5 / California Condor   (*Gymnogyps californianus*)
Only one zoo in the entire world has a specimen of this drab, large bird, which has never been bred in captivity. It is shown here because it is one of the rarest birds in the world. There are only fifty left, and all are found in southern California. Hopefully the educational experience of the zoo will help keep this from happening to other species.

6 / King Vulture   (*Sarcorhamphus papa*)
This bird of the deep tropical forests of Central and South America is the most brilliantly marked of the vultures and a popular zoo exhibit. That bare head helps vultures keep clean. Despite their diet they are very clean animals. This species is black and white with a red, orange, yellow, and black head and neck.

1 / Peafowl   (*Pavo cristatus*)
Few events in the zoo will bring people on the run more readily than the peacock beginning to display for his hens. The fantastic three- to four-foot-long electric green and blue tail with its eye spots doesn't even seem real. It is, though, and happily peafowl, who come from Asia, breed very well in zoos. Peacocks are the male peafowl.

VANISHING ANIMAL

2 / Mikado Pheasant   (*Syrmaticus mikado*)
This splendid, glossy bluish purple bird comes from the island of Taiwan. Although it is rare even there, there are specimens in thirty-five zoos now, and some zoos are having success breeding them. Pheasants are very popular exhibits because of their handsome plumage, their long showy tails, and striking colors. They spend most of their time on the ground and don't fly very often.

3 / Rock Partridge   (*Alectoris graeca*)
This southern European species is only one of the many quails and partridges that are seen in zoos. They are all related to the pheasants and peafowl—a very attractive order of birds from the zoo's point of view. This species, with its distinctive white throat patch, is even found above the tree line in the mountains of Europe and also likes evergreen forest areas.

4 / Vulturine Guinea Fowl   (*Acryllium vulturinum*)
The magnificent cobalt-blue of this bird always attracts comments from zoo visitors, and this beautiful African species can be bred in captivity. In the wild this long-tailed species prefers dry thornbush country and semideserts. It has a shrill, rather unpleasant voice.

5 / Francolin   (*Francolinus* sp.)
In Africa people shoot francolins for the table. In our zoo the francolin is an honored visitor from another continent—an exotic bird. The francolins are related to the quails. They are very chicken-like and are usually found on the ground.

6 / Lady Amherst's Pheasant   (*Chrysolophus amherstiae*)
This magnificent visitor from China and Burma is one of the most beautiful birds in the world. When we look at it, it seems strange to think that it wasn't invented by an artist. It has white, black, gray, red, blue, green, brown, and gold plumage!

1 / Peters' Finfoot (*Podica senegalensis*)
This African water bird is only superficially like a grebe. It has long stiff tail feathers and likes rivers with brushy banks. In a zoo pond it will stay near the edge, particularly if there is growth there. It is duck-sized and swims low in the water. Its call is *keeee*.

2 / Purple Gallinule (*Porphyrula martinica*)
The purple gallinule from the southern United States is one of the most beautiful of all water birds. It is deep purple and bronze-green. It has a yellow frontal shield on its forehead. Its long slender toes distribute the bird's weight and allow it to walk on lily pads. The gallinule is commonly seen in zoos and is a popular attraction.

3 / Sarus Crane (*Grus antigone*)
These handsome gray birds from Asia can stand five feet tall, and they have a red head and upper neck. They make quiet and sensible zoo citizens. You have to watch that beak, though. A poke could result in a terrible eye injury. Keepers know that and are always alert when a sarus crane moves its head.

4 / Greater Sun Bittern (*Eurypyga major*)
It isn't easy to get the bitterns to build a nest and lay eggs in captivity. Occasionally it happens with some species, and it is an important day in the zoo when the young appear. This species is found in Mexico and Central and South America. Its barred and spotted plumage contains many subtle and surprising colors.

5 / Coot (*Fulica* sp.)
Coots are found all around the world, and people often mistake them for ducks. They swim with ducks and look a little like ducks. But look at the coots' feet—not webbed. In many parts of the country they just fly in and join the zoo's waterfowl collection. They are usually a slaty gray with a head and neck that is almost black.

6 / Collared Hemipode (*Pedionomus torquatus*)
A better name for these little Australian busybodies is plain-wanderers. They don't like to fly, and when they are disturbed, they dash around and then freeze, apparently thinking they are safe that way. Their upper parts are brown, their breasts pale buff, and their abdomens white. They don't have any close relatives.

1 / Herring Gull   (*Larus argentatus*)
While some zoos have herring gulls as part of their collection, many don't bother. Near the coast gulls put themselves in the zoo. They will go anywhere there is food, for they have ferocious appetites and, unfortunately, steal eggs. They are found in many parts of the world and are handsome pearly gray bandits.

2 / Puffin   (*Fratercula arctica*)
The puffin looks like a bird that a committee designed—and argued about. The red-tipped, triangular bill is very colorful. The plumage is black and white. In the wild these birds gather in absolutely enormous colonies. Puffins are flying birds and not at all related to penguins. They are native to Iceland and other parts of the North Atlantic.

3 / European Oyster-Catcher   (*Haematopus ostralegus*)
Zoos that have a good shore-bird collection may exhibit this handsome European beach walker. It is seventeen inches long when fully grown and has black and white feathers and a bright coral-red bill. The American oystercatcher looks very much like this bird, who cries *wheep!* as it flies along over the beach.

4 / Kori Bustard   (*Ardeotis kori*)
The bustards walk around open plains areas of Africa and mingle with herds of hoofed animals. Zebras and bustards look very natural together. It is interesting when zoos have an African plains exhibit and let birds like bustards mingle with their natural African neighbors. Most bustards are fairly drably colored—gray or black. Some have chestnut markings, but not this species.

5 / Wilson's Snipe   (*Capella gallinago delicata*)
This bog and marsh bird is related to the sandpipers and woodcocks. It is found over vast areas of eastern Canada and the United States. It is a brown bird with a striped back and a very long, slender bill. It is a favorite target of some hunters.

1 / Crested Pigeon   (*Ocyphaps lophotes*)
This handsome foot-long Australian pigeon is a very pleasant zoo citizen. Pigeons and doves (actually the words can be used interchangeably) are interesting aviary birds since they are usually large and colorful. In the wild this species with its upright feathered crest prefers wooded areas, but not deep forests.

2 / Imperial Pigeon   (*Ducula* sp.)
Many species of pigeons and doves breed without difficulty in captivity, including a few of the imperial pigeons. The imperials are large fruit-eating pigeons from the southwest Pacific world. They lay only one egg when they nest. The blues and greens in their wings are often iridescent and quite beautiful.

3 / Crowned Pigeon   (*Goura cristata*)
The crowned pigeons (there are three species) are also called *gouras*. They are the largest of all the pigeons and doves and can be the size of hen turkeys. This species has bred in zoos and is a very popular attraction because of its great beauty. It is a brilliant blue with some purple and white. It comes from New Guinea and the surrounding area.

4 / Pallas's Sand Grouse   (*Syrrhaptes paradoxus*)
Despite its name, this bird isn't a grouse at all. The sand grouse is related to the pigeon, although not one itself. It comes from central Asia but every now and then starts migrating and may end up in China and England. Like all sand grouse, this species is ground-colored—brown, black, and gray. Males have orange on their necks.

1 / Masked Lovebird   (*Agapornis personata*)
All the true lovebirds come from Africa. This green species with its blackish brown head is from East Africa. Lovebirds are bossy, talkative characters who don't seem to know how small they are—about six inches long. Unless a community cage is very, very large, lovebirds don't do well with other birds because they are too assertive.

2 / Sulphur-crested Cockatoo   (*Cacatua galerita*)
This splendid twenty-inch Australian bird is an important zoo attraction, and some zoos have managed to breed cockatoos. They are white with a yellow crest. During their long life cockatoos develop distinct personalities. They have a harsh voice and come from timbered country, though they also like rivers and farmland.

3 / African Gray Parrot   (*Psittacus erithacus*)
This popular African bird is one of the best "talkers" in the whole world of birds. Parrots don't talk, of course, in the sense of understanding what they say. They just have the ability to mimic sounds. This species is largely gray, but it has a flashy red tail.

4 / Hyacinth Macaw   (*Anodorhyncus hyacinthinus*)
People often call these very large and brilliantly colored birds parrots, and though they are related to the parrots, they are really macaws. All twenty-four kinds of macaws come from Mexico and Central and South America. They are among the most spectacular birds in the world. This species is a brilliant blue, has a light moustache, and is almost three feet long.

5 / Donaldson's Turaco   (*Tauraco leucotis*)
All of the turacos (you can also spell that touraco) are from Africa, and they are fruit-eaters. They are especially attractive birds and do very well in big community cages—the bigger the better. They are quite large (about a foot and a half long) and have harsh voices. They like well-wooded country.

6 / Red-crested Turaco   (*Tauraco erythrolophus*)
An interesting characteristic of the turacos is their special toes that enable them to run and jump about in the trees more readily than most other birds. Look for that behavior the next time you visit the bird house. The bright red crest of this species easily distinguishes it from the others. It is about crow-sized.

1 / Tawny Owl   (*Strix aluco*)
Owls are among the favored zoo exhibits. These birds may live around us but are rarely seen, since they usually hunt at night. This species from Europe and North Africa has black eyes and no feathered ear tufts. It is about fifteen inches long and can be either grayish or brownish.

2 / Barn Owl   (*Tyto alba*)
This is one of the most common of the large birds of prey and is found around the world. (Scientists use the word "cosmopolitan" to describe that situation.) Barn owls are seen in almost all zoos. They are long-legged, and each eye is set in a prominent "facial disc" that gives their face a heart-shaped appearance. There are no ear tufts.

3 / Snowy Owl   (*Nyctea scandiaca*)
The snowy owl of the Far North is one of the most beautiful owls, and, I think, one of the most beautiful birds. It is very white, although the female will have black bands and marks to help conceal her on her ground nest. A great many zoos exhibit a pair of these magnificent birds.

4 / Burrowing Owl   (*Speotyto cunicularia*)
This funny little owl from North, Central, and South America makes its nest underground, in the burrows of other animals. If the zoo provides a burrow for these owls, they threaten the keepers, seemingly unaware that they're only about eight inches tall! They have very long legs for an owl. They are ground-colored and often quite drab.

5 / Tawny Frogmouth   (*Podargus strigoides*)
These very large-mouthed birds come from Australia and usually move around at night. During the day, if they feel themselves in danger, they pose like a broken branch and try to escape notice, as the bird in the drawing is doing. The detail drawing of this gray-brown bird shows its interesting face. The frogmouths are not related to the owls.

1 / Quetzal (*Pharomachrus mocino*)
Sometimes called the resplendent trogon, the quetzal is one of the zoo's most spectacular residents. If you collect stamps, you will recognize this bird as it appears on many Guatemalan stamps. It is brilliantly colored in green and red and keeps to mountain forests where the Aztec and Mayan Indians used to kill it for its tail feathers. It is found from southern Mexico as far south as Costa Rica.

2 / White-tipped Sicklebill (*Eutoxeres aquila*)
Many zoos have special rooms, sometimes called "jewel rooms," in their bird houses for the smallest and most brilliantly colored birds. This electric green and brown gem is found in the wild from Costa Rica to Ecuador. It is five inches long.

3 / Mousebird (*Colius* sp.)
The mousebirds, often called colies, are found in Africa and are about sparrow-size. There are six different kinds of these very sociable birds, and they, too, are jewels that deserve special attention when they appear in the zoo. Their tails have ten feathers and can be ten inches long. Their outer toes are reversible and can point either forward or backward.

4 / Jamaican Mango Hummingbird (*Anthrocothorax mango*)
These tiny little birds perform incredible feats. Some can fly at almost fifty miles an hour, and some beat their wings two hundred times a second. By comparison, a sparrow flaps its wings thirteen times a second. Perhaps the jewel room should be called the gymnasium!

5 / Racket-tailed Hummingbird (*Discosura* sp.)
If you stand by the glass-fronted enclosure in the aviary's jewel room, you will see the tiny hummingbirds hovering like helicopters while they thrust their long beaks into glass tubes. Those tubes contain sugar, honey, water, and other special dietary ingredients hummingbirds need. This species gets its name from its very elongated tail feathers. They look like two very long-handled Ping-Pong paddles.

1 / Turquoise-browed Motmot (*Eumomota superciliosa*)
This lovely bird from Mexico and Costa Rica is an interesting exhibit in an aviary. It is thirteen inches long, largely because of its long tail feathers. Its colors include blue, brown, green, and pink-brown.

2 / Hoopoe (*Upupa epops*)
If you have trouble with scientific names, you can at least have fun with this one! Hoopoes, found in Europe, Africa, and Asia, are dirty birds, famous for having the filthiest nests in birddom. They have the reputation of not being very smart, but they are handsome. They are almost a foot long.

3 / Broad-mouthed Dollarbird (*Eurystomus orientalis*)
This bird belongs to a group called rollers, birds found in Europe, Asia, Africa, and Australia. Most, however, come from Africa. It is a powerful acrobatic flier that actually somersaults and rolls in the air. This species is blue-green with red feet and bill. That's a female coming out of her tree-hole nest.

4 / Blue-cheeked Bee Eater (*Merops superciliosus*)
This bird comes from the Old World—from Africa to New Guinea. All the bee eaters are lovely-looking birds of trim shape, small size (under one foot), and brilliant shades of green and reddish brown. They are bright spots of light in a community cage or aviary.

5 / Blue-breasted Kingfisher (*Halcyon malimbicus*)
There are over eighty kinds of kingfishers, and they are found all over the world. They are colorful, active, and interesting, and they should always be displayed with a water tank of some kind. Kingfishers, it will not surprise you to learn, fish for a living. They are compact, short-tailed, and usually have a crest. Most are some shade of blue, at least in part.

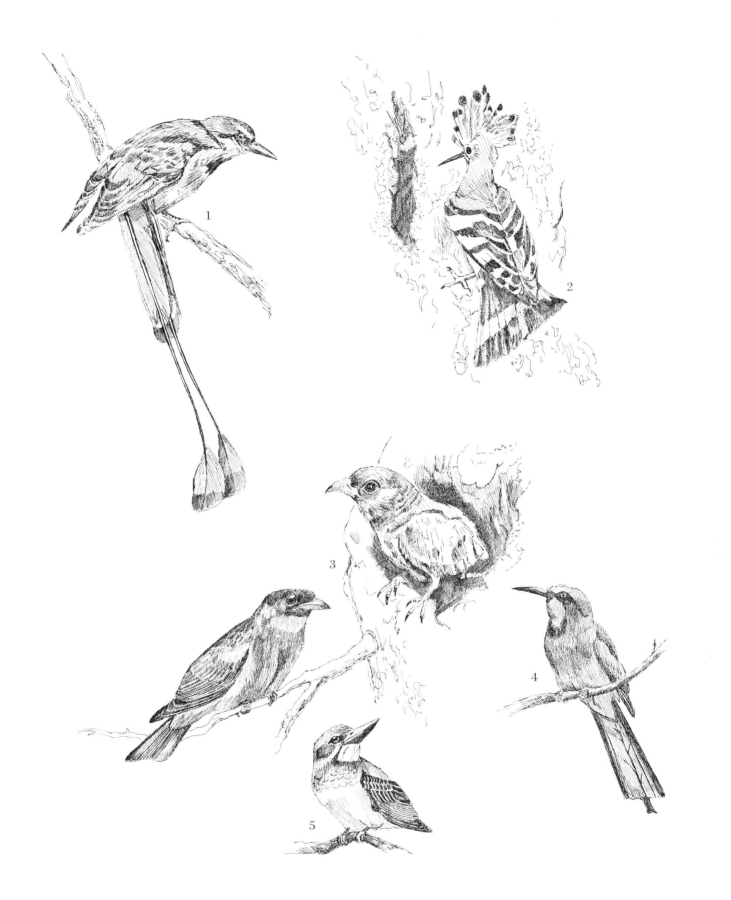

1 / Orange-crested Gardener (*Amblyornis subalaris*)
This incredible bird from southeastern New Guinea is a bowerbird, one of those unbelievable creatures who build a stage to entertain their lady loves and then decorate it! This species fashions a domed arena, with a bank of moss in the middle, and then trims it with flowers that it collects nearby. Some make their constructions in captivity. This nine-inch species is a mottled brown with a bright orange crest.

2 / Greater Bird of Paradise (*Paradisaea apoda*)
This bird belongs to another group of seemingly impossible birds from New Guinea—the spectacular birds of paradise. They are generally deep forest birds and have some of the most spectacular feathers in the world. They are for the male to use to attract the female. Each species has a special way of displaying its finery.

3 / Cock of the Rock (*Rupicola peruviana*)
When you see this crested, intensely colored scarlet and orange bird from South America in the aviary, you will think you are seeing an artist's creation. It surely is one of the most splendid of all birds. The crest is most unusual as it appears to enclose the bill. This species is the most deeply colored—almost scarlet.

4 / Superb Lyrebird (*Menura superba*)
There are two species of lyrebird in eastern Australia, and there is nothing like these birds anywhere else on earth. They are not especially colorful, but the tail of the male (that is the female pecking at the ground) is a great showpiece. As is the case with the peacock and the birds of paradise, the great feathery display is to attract females.

5 / Three-wattled Bellbird (*Procnias tricarunculata*)
If you are lucky enough to be in the bird house when this foot-long chocolate and white bird from Central America sings out, you will not believe your ears. The bellbird sounds like a bell being struck by a metal hammer, and in its native jungle this bird's startling sound can be heard for over a half a mile. Once it starts chiming, even in the zoo, it goes on and on and on.

1 / Green Magpie (*Cissa chinensis*)
The magpies are closely related to the crows, ravens, and jays and like them are sassy and aggressive. This lovely green and brown bird from Asia builds a large and bulky nest and is very bossy with other birds.

2 / Blue Jay (*Cyanocitta cristata*)
If the zoo you are visiting is in native blue jay country (the eastern half of the United States), you won't have to look for these birds in cages. Just look around the grounds, wherever food is to be had. This noisy, pushy bird takes over whenever it has the chance and tells all the other birds what to do. This foot-long bird is, of course, very bright blue and has a prominent crest.

3 / Raven (*Corvus corax*)
The large, glossy black raven is also noisy and very assertive. It can be over two feet long. You will probably see it in an enclosure by itself, for it is too big and too much the bully to stay with many other species. Ravens are among the cleverest birds on earth. They are found in many parts of the world.

4 / Baltimore Oriole (*Icterus galbula*)
These proud natives of North America (they spend the winter in Central America) aren't very big, but their lovely orange and black markings make them among this country's most beautiful birds. The rich and exciting song of this oriole is more a whistle that sounds like *hugh-leeeee*. It likes shade trees, particularly elms.

1 / Ash-throated Flycatcher   (*Myiarchus cinerascens*)
This lovely little American bird, with its pale yellow belly, can be found from the Canadian border all the way down into Mexico. It is a member of the largest group of small birds—the passerines or perching birds. Passerines have four unwebbed toes—three in front and one behind. The hind toe is the strongest and can't be pointed forward.

2 / Rothchild's Mynah   (*Leucopsar rothschildi*)
Over sixty-two zoos now have specimens of this beautiful bird from Bali, and two zoos in the United States are now breeding them. This white bird is now believed to be quite rare in the wild. It has long, thin crest feathers.

3 / Red-eared Bulbul   (*Pycnonotus jocosa*)
This handsome little bird comes from India and Burma. It is another of the passerines you can find in your zoo's community flight cages. Not only is it pleasant to see, but its sweet musical notes are pleasant to listen to. It has a rich brown back and a white, yellow, and black crested head with a splash of red on its cheeks and under its tail.

4 / Melodious Laughing Thrush   (*Garrulax canorus*)
This is an Old World passerine, one of the babblers. As its name will tell you, it has a full, rich voice. It is a moist country bird and will do well in a flight cage planted with vegetation and having a small stream running through it. Its basic colors are brown and white.

5 / Bald Crow   (*Picathartes gymnocephalus*)
These fascinating-looking specimens are not really crows, but exotic birds from Africa. The feathers on their necks are so fine and smooth that they seem like leather or satin. Although they are becoming rare in the wild, fourteen zoos have groups of these unusual birds. They are grayish brown and white.

1 / Loggerhead Shrike   (*Lanius ludovicianus*)
This smallish perching bird is found in the eastern parts of the United States and Canada. It is gray above and white below and has a black bandit's mask across its eyes. It has a large head and a rather long tail.

2 / Paradise Whidah   (*Vidua paradisaea*)
The whidahs are also called widow birds, and they come from Africa. The four middle feathers of the male's tail are longer than the rest and form a handsome train. This species is grayish black on top with a brownish underside.

3 / Scarlet Tanager   (*Piranga olivacea*)
This native American bird is a star in any zoo exhibit, and the male is much more colorful than the female, as is true in most every kind of bird. Scientists call this marked difference in the appearance of the sexes *sexual dimorphism*. The male in this species is scarlet with black wings and tail. The female is a dull green above and yellowish below with blackish or brownish wings.

4 / Purple-rumped Sunbird   (*Nectarina zeylonica*)
Hummingbirds come from the New World, and their counterparts in the Old World are the gorgeous sunbirds. This one comes from India and Ceylon. It is interesting to note that its scientific name tells you what it eats—nectar—and where it is found. This incredible species is metallic crimson, green, purple, and yellow. Even the females are lovely.

5 / Golden Sunbird   (*Anthreptes* sp.)
The golden sunbird from Africa is a stunning example of this beautiful group of tiny birds whose brilliant plumage is designed to camoflage the bird as it flies among the flowers. The bright feathers of the male also play an important role in attracting the females during the mating season, thus insuring continued reproduction of the species.

6 / European Goldfinch   (*Carduelia carduelis*)
The goldfinch is further proof that we don't have to rush off to exotic lands to find beauty. Just across the ocean in Europe these beautiful birds are very common. They have wide yellow wing bars and are black, brown, and white. The male has a very red face. The call is an often repeated *deedelit*.

VANISHING ANIMAL

1 / Galapagos Tortoise   (*Testudo elephantopus*)
The mighty tortoises of the Galapagos Islands are true giants, weighing several hundred pounds, and all of them are unfortunately rare. There has been some small success with breeding them in captivity, but not nearly enough. They are quiet, gentle animals and popular zoo attractions. Tortoises require good care and a great deal of green food every day.

2 / Florida Box Turtle   (*Terrapene carolina bauri*)
Compare this five-and-a-half-inch turtle with the giant from Galapagos! Turtles and tortoises come in a variety of sizes and are interesting members of the reptile house community. It is sad how many of these animals are becoming rare. Uncontrolled cats and dogs kill many, as do cars. The destruction of the turtles' habitats and pollution are also contributing factors. They are ancient, peaceful animals, and it would be a great shame to lose them through our own carelessness.

3 / Atlantic Green Turtle   (*Chelonia mydas mydas*)
This three-foot-long turtle inhabits the warmer parts of the Atlantic Ocean, the Gulf of Mexico, and the Mediterranean Sea. Since sea turtles are reptiles, they are cold-blooded, and temperature is very important to them. All sea turtles make popular exhibits in aquariums, but not every zoo and aquarium can afford the time, space, and money needed to maintain such animals in captivity. Those without proper facilities shouldn't even try.

1 / Nile Crocodile   (*Crocodylus niloticus*)
Seeing a fifteen- or sixteen-foot crocodile anywhere can be an exciting event. Probably because there are so many frightening stories told about these ancient animals, people stare at them with wonder and fascination. They are egg-laying reptiles and very, very ancient. Long before nature even got the idea of creating a dinasaur, the crocodiles were on earth and doing fine. Now the dinosaurs are gone, but the mighty, almost legendary crocodiles are still with us. That makes them even more fascinating.

2 / Florida Alligator   (*Alligator mississippiensis*)
Alligators make another excellent exhibit, and they do well in zoos if properly cared for. They are found only in North America and China, strangely enough, and some zoo collections have both kinds. Two American zoos have had success in breeding alligators in captivity, and that is at least a start. We had better do something. Alligators, like all of the crocodilians, are becoming increasingly rare in the wild.

3 / Gavial   (*Gavialis gangeticus*)
Most people know alligators and crocodiles, but a surprising number don't know about the slender-snouted gavials. Like all their cousins in the crocodile family, the gavials have been hurt by overhunting. They are killed for their skins and sometimes out of plain ignorance. There are gavials in a number of zoos now, but nobody seems to be having much luck breeding them. They come from several of India's vast river systems.

1 / Jackson's Chameleon   (*Chamaeleo jacksoni*)
These interesting African lizards are not easy to maintain in captivity. They require just the right temperature and just the right amount of moisture. They also must have the right bugs to eat. They are so plentiful in Africa that zoos there buy them for about a penny apiece and use them to feed the captive snakes. We treat them differently here, as though they are rather exciting, exotic animals, and that is exactly what they are. They give birth to living young, and often chameleons obtained for a zoo will be carrying young, though it isn't an easy task to keep the babies alive once they are born.

2 / Komodo Dragon   (*Varanus komodoensis*)
This giant from the East Indies lives on only one single island—Komodo. This is the largest of the monitor lizards and can reach a length of ten feet and a weight of three hundred pounds. In the wild, these incredible giants hunt pigs and deer. There are a number of specimens in zoos around the world, but no one seems to be breeding them. Seeing any true giant is exciting, and the Komodo dragon lizard is near the top of the list. Compare this animal with some of the tiny lizards you see in the reptile house.

3 / Cuban Ground Iguana   (*Cyclura macleayi*)
The great lizard family Iguanidae contains about seven hundred species, and most come from the Western Hemisphere. All the iguanas we know in the zoo do, and they can be up to five feet long in some species. Perhaps one of the reasons they are so popular with zoo goers is that they look somewhat similar to what we might imagine a dragon would look—breathing fire and all. The iguanas get a little touchy when they are grown, and many of them bite. But, in the zoo, they make good, sensible residents.

4 / Gila Monster   (*Heloderma suspectum*)
Despite all the silly stories, there are only two poisonous lizards in the entire world, and they come from Mexico and the southwestern United States. One is the Gila monster pictured on the opposite page, and the other its close cousin, the Mexican beaded lizard. They are shy in the wild and certainly don't go around attacking people. Still, they should never be handled by anyone but an expert. The Gila monster is not an animal for the amateur to care for.

VANISHING ANIMAL

1 / Indian Python   (*Python molurus*)
This is one of the six longest snakes in the world and does grow to be twenty feet from nose to tail. Snakes that long are very thick, of course, and some people think they are man-eaters. Pythons do not go around eating people, or strangling people, or anything of the sort. The next time you are in the reptile house watch the people who look at the pythons. You can tell from their faces that their imaginations are working overtime. That's what the magic gateway of the zoo is all about. It gives everyone a chance to go on safari.

2 / Boa Constrictor   (*Boa constrictor*)
There are no pythons in the Western Hemisphere, but there are boa constrictors. (Notice that the common name and the scientific name of this animal are the same. I don't know of any other case where this is true.) Specimens of the boa can get to be almost nineteen feet long, but certainly not the nonsense lengths of one hundred feet we hear about in adventure tales. Boas are active hunters, but they don't hunt men!

3 / Indigo Snake   (*Drymarchon corais*)
This beautiful and harmless snake is found in Mexico and southern Texas. It is one of the gentlest of all snakes. Even fully-matured specimens caught in the wild can be expected to settle down in a day or two and allow themselves to be handled without protesting. Of course, it is better not to catch such snakes, but allow them to stay wild where they will breed and hunt rodents. In the zoo they serve an educational purpose.

4 / The Black-necked Garter Snake   (*Thamnophis cyrtopsis*)
There are many forms of garter snakes and a great many color variations. Basically, they are harmless, slender, often colorful snakes with not a very nice disposition. They will bite when handled, but the bite is completely harmless. They are often seen in zoos, although they aren't what you would call an important exhibit. They do become important when the exhibit is designed to show the harmless snakes of the region and explain the good they do. Garter snakes should never be killed or even molested in the wild.

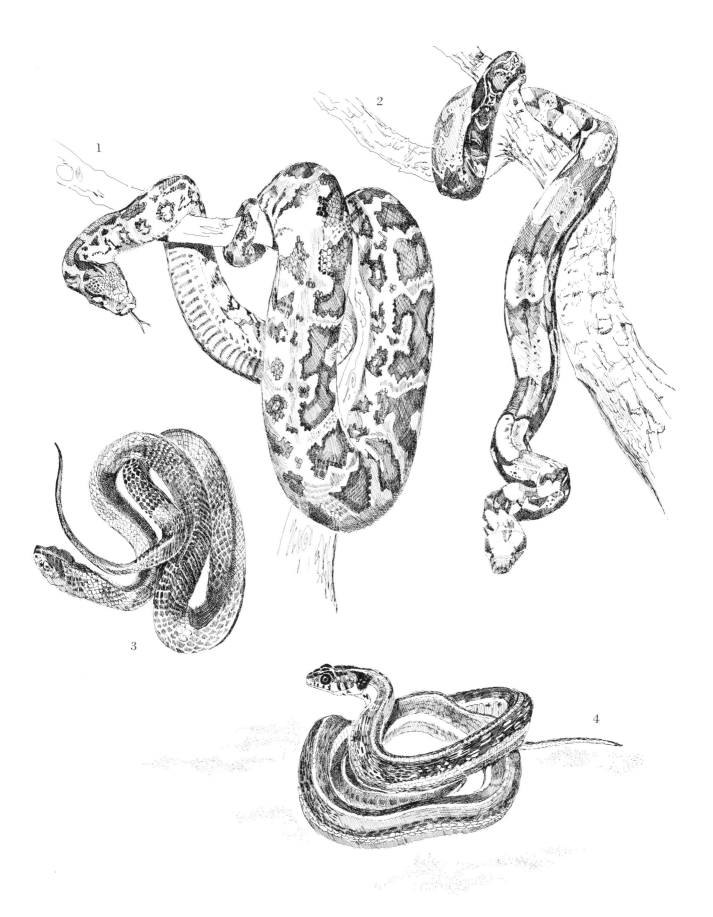

1 / Gaboon Viper (*Bitis gabonica*)
This powerful poisonous snake grows to be six feet long and has what are probably the longest snake fangs in the world—almost two inches. The Gaboon viper comes from Africa, and although it is dangerously poisonous, it is quite docile. Keepers in the reptile house don't take chances, however—not with poisonous snakes with two-inch fangs! (Note the detail drawing of the head.)

2 / Eastern Diamondback Rattlesnake (*Crotalus adamanteus*)
This is one of the largest and most dangerous of the rattlesnakes. Very large specimens can be almost eight feet long, although that is a true giant. Rattlesnakes are alert snakes and can be quick to strike. Working with animals like these in the zoo requires special training and long experience. The cages are padlocked in back so that there are no escapes. Snakes, generally, don't want to escape. Their cage is their world, and all their needs are supplied—heat, water, food, safety. A snake doesn't need or want exercise.

3 / Indian Cobra (*Naja naja*)
What would the zoo be without its cobras? Everyone expects to see them in the reptile house, and visitors are rarely disappointed. Cobras are very common in Africa and Asia and are of course venomous, but not to the degree that some people think. Most people bitten by cobras manage to survive. You should *never* wave your hand in front of a snake enclosure or tap on the glass to make the snake strike. Snakes striking the glass (which they apparently can't see) injure themselves, and those injuries can cause death.

1 / American Toad   (*Bufo americanus*)
Toads can make interesting small exhibits, and many, in fact, are seen displayed in reptile houses. They aren't reptiles, of course, but amphibians, as are all the animals on the opposite page. There is no truth to the story that toads give you warts. What toads do give you are fewer insects. They consume great quantities of them and help make our lives more pleasant.

2 / Caecilian   (*Schistometopum* sp.)
The caecilians are small burrowing amphibians that live in soft earth and are seldom seen by man. There are about seventy-five species, and they are very remote from our lives. Occasionally a zoo will have special exhibits of animals like these, very often in the reptile house where there are numerous small glass enclosures. It is fascinating to see the diverse forms life can take. The caecilian may seem strange to us, but it is, after all, a backboned animal.

3 / Marsh Frog   (*Rana* sp.)
Frogs, like toads, are the familiar animals of the world of the amphibians. They are the animals we know. Many kinds of frogs are exhibited in zoo collections. The combined study of reptiles and amphibians is known as herpetology—and the animals of these groups seen in zoos are known by the nickname *herps*. Frogs are often colorful and can be quite comical to watch. It is most interesting to watch their young develop from tadpoles to adults.

4 / Yellow-blotched Salamander   (*Ensatina croceator*)
Salamanders range in size from an inch or so to five feet. This little species comes from the coastal mountains of southern California. Salamanders don't make very exciting exhibits when you compare them to cobras and pythons, but they are interesting and often quite beautiful. Unlike snakes, salamanders are slimy to the touch.